Brain Theory

Proceedings of the First Trieste Meeting on Brain Theory, October 1–4, 1984

Edited by
Günther Palm and Ad Aertsen

With 75 Figures

Springer-Verlag Berlin
Heidelberg GmbH

Dr. GÜNTHER PALM
Dr. AD AERTSEN
Max-Planck-Institut für
Biologische Kybernetik
Spemannstraße 38
7400 Tübingen, FRG

Cover illustration:
Reactive and creative system differing in absence or presence of closed loops. From P. Johannesma, this Volume, page 34, Figure 6.

ISBN 978-3-642-70913-5 ISBN 978-3-642-70911-1 (eBook)
DOI 10.1007/978-3-642-70911-1

Library of Congress Cataloging in Publication Data. Trieste Meeting on Brain Theory (1st : 1984) Brain theory. 1. Brain–Congresses. 2. Intellect–Congresses. 3. Neurology–Philosophy–Congresses. I. Palm, Günther, 1949– . II. Aertsen, Adrianus, 1948– III. Title. IV. Title: Brain theory. QP376.T73 1984 153 85-27664

Softcover reprint of the hardcover 1st edition 1986

Offsetprinting: Beltz Offsetdruck, Hemsbach/Bergstraße.

2131/3130-543210

Preface

The present collection of papers forms the Proceedings of the First Meeting on Brain Theory, held October 1–4, 1984 at the International Centre for Theoretical Physics in Trieste, Italy.

The Meeting was organized with the aim of bringing together brain theorists who are willing to put their own research in the perspective of the general development of neuroscience. Such a meeting was considered necessary since the explosion of experimental work in neuroscience during the last decades has not been accompanied by an adequate development on the theoretical side.

The intensity of the discussions during the Meeting is probably reflected best in the report of the organizers, reprinted here following the Preface.

During the Meeting it was decided that a workshop of this kind should be repeated at regular intervals of approximately 2 years. The International Centre for Theoretical Physics in Trieste has kindly agreed to act as host for future meetings.

The present Meeting was supported by grants from the International Centre for Theoretical Physics and the International School for Advanced Studies in Trieste, IBM-Germany through the "Stifterverband für die Deutsche Wissenschaft" and the Max-Planck-Institute for Biological Cybernetics.

G. PALM
A. AERTSEN

Preface

The present collection of papers forms the Proceedings of the First Meeting on Brain Theory, held October 1–4, 1984 at the International Centre for Theoretical Physics in Trieste, Italy.

The Meeting was organized with the aim of bringing together brain theorists who are willing to put their own research in the perspective of the general development of neuroscience. Such a meeting was considered necessary since the explosion of experimental work in neuroscience during the last decades has not been accompanied by an adequate development on the theoretical side.

The intensity of the discussions during the Meeting is probably reflected best in the report of the organizers, reprinted here following the Preface.

During the Meeting it was decided that a workshop of this kind should be repeated at regular intervals of approximately 2 years. The International Centre for Theoretical Physics in Trieste has kindly agreed to act as host for future meetings.

The present Meeting was supported by grants from the International Centre for Theoretical Physics and the International School for Advanced Studies in Trieste, IBM-Germany through the "Stifterverband für die Deutsche Wissenschaft" and the Max-Planck-Institute for Biological Cybernetics.

G. Palm
A. Aertsen

Contents

Reviews of Historical Papers

List of Contributors

You will find the addresses at the beginning of the respective contributions

Report of the First Meeting on Brain Theory

held at the International Centre for Theoretical Physics in Miramare, (Trieste) Italy, October 1–4, 1984

V. BRAITENBERG and G. PALM[1]

The meeting was, in the opinion expressed by all participants, a very successful one. Each of the presentations elicited strong responses in the majority of the participants leading to an unusually vivacious and detailed discussion in which not only common ideas emerged in unexpected ways but also weak or obscure points of the individual theories where freely discussed and clarified. It soon became obvious that we had created a forum which provided an opportunity to discuss ideas which were often confined to a strange limbus of awe and contempt in previous discussions with experimental neuroscientists. A consequence of this realization was probably the absence of the aggressively competitive spirit that can easily emerge in similar groups. We attributed this in part to the fact that no representative of the staunch type of experimentalist was present at the meeting. Thus we could make uninhibited use of mathematical formalism and propose ideas that are not directly related to experimental research presently en vogue, but may well provide the framework for future experimentation. This is to be expected since at our meeting it became apparent that the goal of finding a common theoretical framework is well within reach. Especially in the discussion of the historical papers, for which the last day of the meeting was reserved, it became clear that the quite different views on the brain that are naturally held by scientists from different backgrounds can possibly be transformed into each other on the level of their mathematical representations.

The elements of a mathematical representation of brain theories were already laid out in 1961 in a paper by Caianiello. In his talk he showed the possibilities of an analytical treatment of his equations, which put an extreme emphasis on the nonlinearity of neural responses, and were originally designed as an alternative to the logical formalism prevalent in the early years of brain theory. The ideas of cell assemblies and associative memory, which introduce the necessary teleological argumentation into the pure physics of neural networks, were at the center of the talks by Palm, von der Malsburg, Lansner and Shaw. Palm used Information Theory to develop a quantitative optimization criterion (the information storage capacity) for associative memories and

[1] Max Planck Institut für Biologische Kybernetik, Spemannstraße 38, 7400 Tübingen, FRG

Brain Theory
Edited by G. Palm and A. Aertsen

discussed some results of this optimization. By way of application he also discussed some anatomical data on the cortico-cortical long-range connections of the mouse. Von der Malsburg emphasized the problem of representation of things and their relations in the brain, whether by means of single neurons or assemblies. If things are represented by assemblies, then relations could be represented by connections. This idea leads to the supposition of an unusual synaptic short term memory mechanism, which is similar to Hebb's rule, but on a much faster time scale. The working of this rule was illustrated by some computer simulations. Lansner proposed a hierarchical network structure, for the representation of relations. He is currently performing computer simulations of this structure. Shaw presented a simplified physical model, in the spirit of the Ising Model, where whole assemblies were represented by simple elements (Trions). This model showed remarkably complex behaviour with only few elements.

The other contributions were more closely related to particular aspects of current experimental brain science. Pellionisz' tensor theory is dominated by the ideas of transforming covariant coordinates into contravariant coordinates for skewed coordinate systems, which is obviously involved in the coordination of movements on the basis of proprioceptive, vestibular and other sensory input. He presented a detailed model of the cerebellum, which is based on these ideas and illuminates many anatomical and physiological features. Von Seelen showed how information derived from cortical histology can be made into a dynamic model displaying many features that can be related to physiological experiments. For example, he demonstrated, how the receptive field of a neuron can change with time. This is due to the dominating positive feedback in the network. Freeman gave an exposition of the mammalian olfactory system illuminating a wealth of anatomical and physiological facts, many from his own laboratory, with a variety of theoretical ideas. His main point was the detection of changes in network dynamics and structure in the course of learning, which is evident in EEG potentials. The main point in Braitenberg's talk was that even very elaborate sets of experimental findings leave ample space for constructive speculation. He showed that a simple assumption on the distribution of inhibitory cells in Area 17 allows to deduce many of the effects discovered by Hubel and Wiesel. Aertsen addressed the problem of the digestion of a plethora of data derived from multi-unit recordings. He presented a new method of correlation analysis for such data. The idea is to set up distances between units in a multidimensional space by the action of something like a gravitational force, the strength of which reflects their mutual correlation. Johannesma introduced concepts for the analysis of neural responses to external stimuli (the pre- or peri-spike stimulus ensemble), which can be used for an objective definition of receptive fields. These concepts were illustrated by many examples from his own research in acoustic physiology.

A general consensus was obtained that a similar meeting in about 2 years' time could be expected to become again very fruitful and first organizational steps were taken. The sponsorship of various institutions which provided financial support (International Center for Theoretical Physics, International School for Advanced Studies in Trieste, IBM-Germany through the Stifterverband für die Deutsche Wissenschaft, the Max-Planck-Institut für Biologische Kybernetik and especially those participants who contributed to their travel expenses through their own funds) is gratefully acknowledged.

Introduction

G. PALM and A. AERTSEN[1]

The working of the brain certainly presents one of the most challenging problems in natural science. There is an overwhelming amount of experimental data on the brains of various animals. However, there is no unified brain theory yet. Up to now there exist a number of different theoretical approaches to the understanding of the brain, which originate from different experimental or experiential backgrounds. In some of them this background is quite explicit e.g., in the scope of the theory; others have arrived at more general theories in which the original background remains implicit. This situation is not hopeless, though, since there is at least a common language in which the relevant questions can be discussed, namely that of mathematical formalism.

In a sense, different theories of the brain can be compared to different spectacles through which the brain is observed. A meeting of brain theorists then has the objective of comparing these spectacles. Therefore the participants of such a meeting have to be able and willing to understand the working of each other's spectacles and even to put them on to have a look through them. The presentation of experimental data in such a meeting primarily serves the purpose of illustrating the working of the spectacles by showing a view through them. Usually the emphasis in interpretation of data in neuroscience is on the view, the spectacles through which the view was obtained remain implicit.

The present Proceedings of the First Trieste Meeting on Brain Theory can be regarded as a collection of all those spectacles that were presented and probed during the meeting. This collection contains old spectacles, the reviews of influential theoretical papers (pp. 229–259), and new spectacles, the contributions of the participants (pp. 1–228), many of them improvements and modifications of older ones.

[1] Max Planck Institut für Biologische Kybernetik, Spemannstraße 38, 7400 Tübingen, FRG

Brain Theory
Edited by G. Palm and A. Aertsen

From Neuron to Assembly: Neuronal Organization and Stimulus Representation

A. AERTSEN[1], G. GERSTEIN[2], and P. JOHANNESMA[3]

1 Introduction

The study of information processing in the sensory nervous system may be viewed as an investigation of images. Let us consider, for instance, the auditory nervous system. Throughout the auditory system, starting at the hair cells in the cochlea and the auditory nerve fibres, through the various stages of the auditory processor, composed of the numerous individual neurons with their different patterns of interconnections, we have what might be called "the neural image of sound" in its different realizations. The external world is paralleled by an internal representation (e.g., Craik 1943, McCulloch 1965). The acoustic environment of an animal, consisting of patterns of air pressure variations at the external ears, is represented and transformed internally by a network of neurons which communicate by complex spatio-temporal patterns of action potentials, the all-or-none events generated by the individual neurons.

1.2 Single Unit Analysis

A central paradigm in the study of the sensory nervous system is that meaningful information regarding its principles of operation can be obtained from experimental investigation of the functional characteristics of its elementary components, i.e., the single neurons. This is reflected in the vast amount of literature on single unit experiments. In the periphery, the single neurons can quite adequately be described by characteristics such as frequency tuning and temporal pattern of firing to stimulation with tones. This type of characteristics, grossly speaking, serves to cover all relevant aspects of the various possible sounds occurring in the external world (e.g., Evans 1974). Moreover the periph-

[1] Max-Planck-Institut für Biologische Kybernetik, Spemannstraße 38, 7400 Tübingen, FRG

[2] Department of Physiology, School of Medicine, University of Pennsylvania, Philadelphia, PA 19104, USA

[3] Department of Medical Physics and Biophysics, University of Nijmegen, Nijmegen, The Netherlands

Brain Theory
Edited by G. Palm and A. Aertsen

eral neural image can be characterized by simple organization principles like tonotopy, i.e., an orderly spatial representation of the spectral composition of the acoustic environment. The picture becomes increasingly more complex and shows more and more blanks when progressing towards the more central parts of the auditory system. Classical experiments using "simple" stimuli such as tones and clicks appear to lose their claim to completeness.

A theoretical description and related experimental and data-analysis procedures developed by the Nijmegen group provide for single neurons in more central parts of the auditory nervous system, a functional representation of the neuron as an element of the auditory processor (for a review see Eggermont et al. 1983b). This approach is based on statistical analysis of the relation between the extra-cellularly recorded single unit activity and the presented ensemble of sound stimuli. A question of special interest has been to what extent such a description covers the neuron's behaviour under a variety of different acoustic stimulus conditions (e.g., tones, noise, natural sounds).

The analysis is focused on the evaluation of the spectro-temporal properties of those stimuli which precede the occurrence of action potentials, the neural events, as compared to the characteristics of the complete ensemble of stimuli presented to the animal. This approach, formally related to the evaluation of the second order stimulus-event cross correlation function, results in the "spectro-temporal sensitivity" (STS) as a characteristic of the neuron's stimulus-event relation. In order to get an impression of the extent of stimulus-invariance of this neuron characteristic, the outcome of the correlation analysis has to be normalized with respect to the a priori known spectro-temporal structure of the stimulus ensemble, used to measure it. Under certain conditions, regarding both the neuron's system function and the stimulus ensemble, a stimulus-normalization procedure can indeed be derived using the formalism of nonlinear systems theory. If the spectro-temporal sensitivity of the neuron can be normalized in this way, a "spectro-temporal receptive field" (STRF) can be derived. This STRF forms the functional characteristic of the contribution of this particular neuron to the auditory processor.

Experimental investigations have been made of the receptive field properties of single neurons in the auditory midbrain of the grassfrog (Rana temporaria L.) under a variety of stimulus conditions (tones, noise, species specific vocalizations). It appeared that, even after elaborate stimulus normalization procedures in a considerable number of cases the result cannot be reconciled into a single model of the sinle unit receptive field. This is illustrated in Fig. 1 which shows the spectro-temporal sensitivity for seven neurons from the frog's midbrain, determined with three different types of stimulus ensembles. As a general conclusion it can be stated that in a considerable number of cases the single-unit receptive field turns out to be stimulus-variant, possibly context-sensitive. This, obviously, poses interesting problems when using the receptive field as a conceptual tool in modelling brain function.

Theoretical considerations at the same time have shown that for neuron models which can be characterized as not having any feedback connections

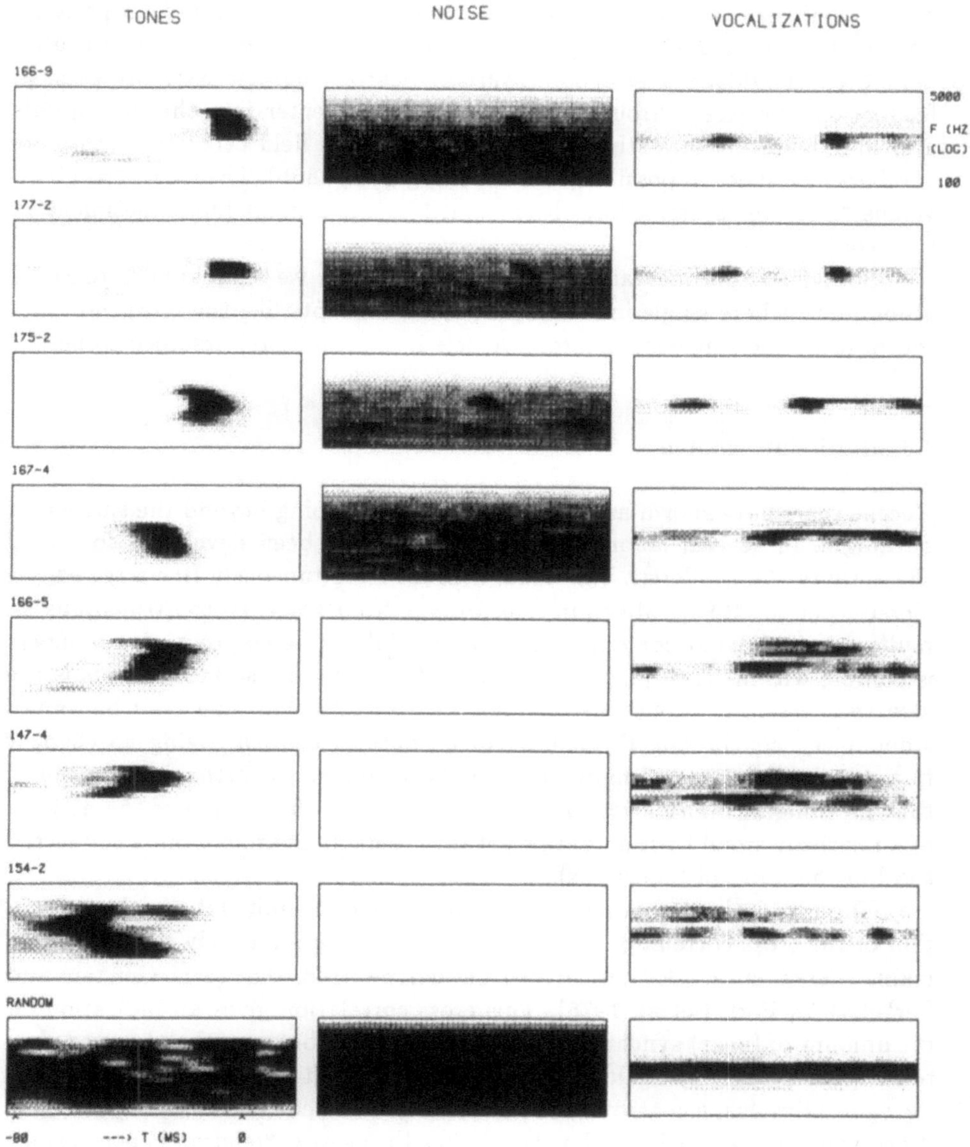

Fig. 1. Spectro-temporal sensitivity of 7 neurons (*upper 7 rows*) in the auditory midbrain of the grassfrog (*Rana temporaria L.*), determined with different types of acoustic stimuli (*3 columns*: tone pips, stationary Gaussian wideband noise and species-specific vocalizations). The average spectro-temporal structure of each of the three stimulus ensembles is shown in the *bottom row*. Further explanation in text. (From Johannesma and Eggermont 1983)

across nonlinear elements, the receptive field can be defined and, in principle be measured in a neat way (Van den Boogaard et al. 1985a). This receptive field then should be a stimulus invariant neuron characteristic. As soon as feedback loops (e.g., through circuits of neurons) enter into the description, a formal definition of a stimulus invariant receptive field becomes much more cumbersome, if not impossible (Van den Boogaard 1985b). The terms "reactive" versus "creative" systems have been coined in this context (Johannesma et al. this Vol.).

Both for experimental and theoretical reasons we thus appear to reach a boundary where single unit analysis, using concepts like the receptive field, ceases to be the single explanatory vehicle it is (implicitly) assumed to be.

1.3 Multi-Unit Analyis

Recent years have shown an increasing interest in going beyond the single unit paradigm. In several laboratories techniques have been developed to record the activity from different individual neurons simultaneously (for a review see Gerstein et al. 1983). Multi-unit recordings from some 10 to 20 neurons simultaneously have become feasible; the possibility of going to higher numbers is coming within reach (e.g., Grinvald 1984). Behind these experiments is the idea that essential properties of single neuron behaviour can only be understood when considering the neuron in its natural context: being an element in a massively connected network of similar elements. A further assumption is that meaningful information can be obtained from observation of the activity of a relatively small fraction of this network (e.g., 10 neurons from a connected total in the order of thousands).

The classical approach to the analysis of multi-unit data is the evaluation of the cross correlation function of pairs of simultaneously recorded spike trains (Perkel et al. 1967), or procedures derived from that (e.g., Gerstein and Perkel 1972, Perkel et al. 1975). This cross correlation gives an indication on the amount of (near) synchrony in firing (or silence) between the observed neurons. Since synchrony could also be induced by both neurons being driven by the same stimulus, a control measurement is performed. The stimulus ensemble is presented at least twice and one evaluates the cross correlation between spike trains from different units, now, however, with the time shift in the correlogram around the stimulus period ("shift predictor"). This correlogram measures the direct stimulus influence on the firing synchrony. The "difference" (not necessarily the algebraic difference) between these two measures of coincidence in firing is interpreted as the sign of true connectivity between the neurons.

Recent results using this type of approach (Eggermont et al. 1983a, Frostig et al. 1983, Bloom and Gerstein 1984, Espinosa and Gerstein 1984) have suggested that a similar problem exists as described above for the single unit receptive field: in a considerable number of cases the neural correlation, after application of a stimulus normalization procedure, may be stimulus-dependent. An example of this is given in Fig. 2, which shows the simultaneous and the non-simultaneous cross correlation of firing for a pair of neurons from the grassfrog's auditory midbrain, determined with different types of stimuli.

This observation gives rise to the hypothesis that the stimulus-variance of both the single unit receptive field and the neural correlation are intrinsically interconnected, in fact they are both manifestations of the same underlying mechanism: the neurons being elements in an interacting population, possibly giving rise to neuronal assemblies or related concepts (Hebb 1949, Braitenberg 1977, Palm 1982). This hypothesis can only be addressed by analysis methods which transcend the usual approach of addressing a group of neurons as a collection of all possible pairs or triplets, but instead, stress the properties of the entire group of neurons as an entity.

Also for purely pragmatic reasons such an approach seems inevitable. The very proliferation of correlation functions resulting from a multi-unit experiment clearly points to the necessity of a more integrated representation of the relevant interactions. A 10-unit recording, for example, results in 45 different cross correlograms to be calculated and, what is more, to be interpreted; a 20-unit recording leaves the poor experimenter with 190 correlograms. Fur-

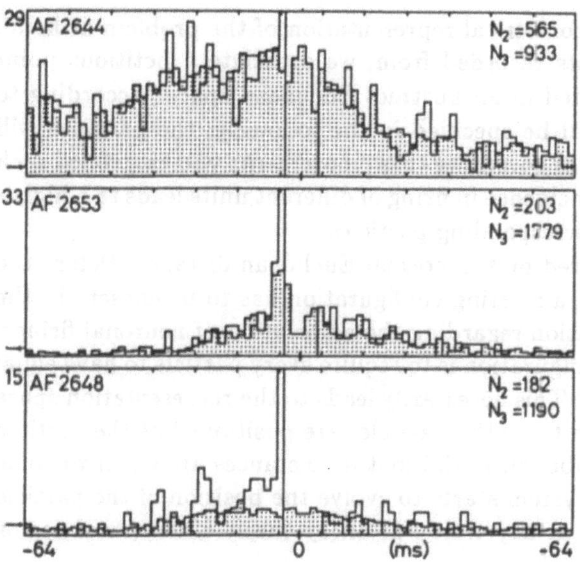

Fig. 2. Simultaneous and non-simultaneous cross coincidence histograms for two neurons from the auditory midbrain of the grassfrog (*Rana temporaria L.*), under tonal and noise stimulation. The non-simultaneous coincidence histograms are shaded. The difference between both histograms is interpreted as a measure for the strength of neural interaction. It is observed that this difference is negligible for stimulation with 48 ms tonepips presented once per s (*upper*); is confined to a few bins in case stimulation was done with 16 ms tonepips presented once per 128 ms (*middle*); and is quite clear and more extended for stimulation with stationary wideband noise (*lower*). (From Eggermont et al. 1983a)

thermore, these numbers have to be multiplied with the numbers of different stimulus conditions and possible other parameters of interest, such as time resolution of the correlogram. A more global and integrative method of analysis seems imperative, if only for logistics reasons.

A new conceptual representation of cooperative behaviour in a population of observed neurons, evolved from earlier ideas in statistical pattern recognition (Wright 1977), has recently been described (Gerstein et al. 1985, Gerstein and Aertsen 1985). This representation leads to a new technique for detecting and studying functional assemblies: "gravitational clustering". The multi-unit problem is mapped into an N-body problem; functional similarity between spike trains is translated into metrical distances. This representation being purely a mental construct, the rules governing the transformation can be defined at will, preferably such that the behaviour of the "model" system in a natural way induces inference regarding the relevant relations in the original problem.

2 Gravitational Representation of Multi-Unit Activity

2.1 Particles, Space, and Charges

As a first step we set up a geometrical representation of the problem at hand. With each one of the N units recorded from, we associate a fictitious point particle; this particle is located in an abstract Euclidean space. According to dynamic equations which will be specified in the following, the particles will travel through space. The rules governing these movements will be defined such that: higher than average coincidence in firing of different units leads to a higher degree of proximity of the corresponding particles.

The proximity is measured by the normal Euclidean distance. Before the particles are set into motion, a starting configuration has to be chosen. In the absence of any prior information regarding the association of neuronal firings, the most natural starting configuration is to require every particle to have equal distance to all the other ones. This necessarily leads to the representation space being N-dimensional: at time $t = 0$ the particles are positioned at the vertices of an N-dimensional hypercube, all initial mutual distances are equal to some starting value d_0. After the system starts to evolve the position of the particle corresponding to neuron i is at any time t given by the N-dimensional vector $\mathbf{r}_i(t)$.

We now have to specify the rules according to which the particles will move. To this end we associate with each particle an "electric charge" $Q_i(t)$, and have this charge be determined by the spike sequence $z_i(t)$ of the corresponding neuron i. In the present work we chose for the charge function a low-pass filtered version of the actual spike sequence:

$$Q_i(t) = \int ds\, q(s) z_i(t-s) \tag{1}$$

with the impulse response $q(s)$ given by

$$q(s) = q_0 \exp(-s/\tau) \ . \tag{2}$$

The time constant τ is a parameter, the value of which is chosen by the investigator. The charge history for a segment of a typical spike sequence is shown in Fig. 3.

Due to the electric charges the particles will exert forces onto each other, and, as a consequence, will start to move. In the present formulation we define the force between any two particles to be proportional to the product of their charges. The direction of the force is along the line conjoining the instantaneous position of the two particles involved, being attractive for equal signs of charge. Analogous to the Coulomb interaction in physics we may, in addition, have the force be distance dependent. The total force acting on a particle is obtained by vectorial addition of the individual contributions from all pair interactions involved.

Note that, unlike the normal case in physics, the electric charges are fluctuating functions of time. Through this time dependence the force field between particles is modulated in time by the activity patterns of the corresponding neurons. As a result those particles that correspond to neurons that tend to fire in synchrony will exert mutual attractive forces, and, consequently those particles will start to aggregate. As time proceeds their mutual distances will keep decreasing; after a sufficiently long interval has passed the sets of synchronously firing neurons will have collapsed to clusters in N-space, with every cluster corresponding to a different coherently firing group of neurons among the population that was recorded from. These clusters can be identified and characterized by standard methods of cluster analysis and pattern recognition (e.g., Fukunaga 1972).

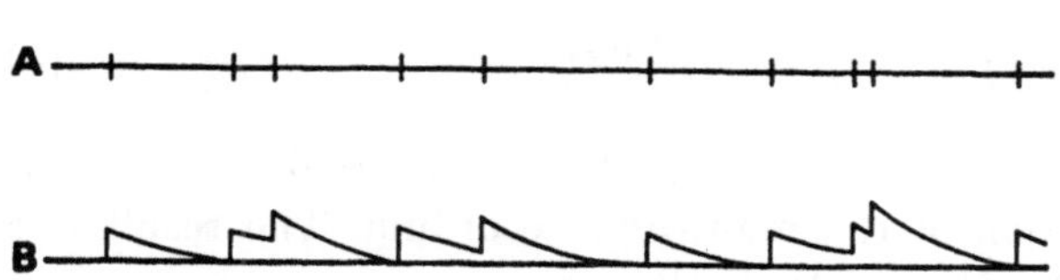

Fig. 3. Conversion of spike train to charge used in the gravitational representation. At the time of each neural impulse, the existing charge of the particle corresponding to that neuron is incremented by a fixed amount. Between neural impulses the charge decays with a fixed time constant. In the "rate normalization" used in the present paper the magnitude of the charge increment is different for each neuron, and is inversely related to its mean firing rate. (From Gerstein et al. 1985)

2.2 Dynamic Equations

The propulsive field $\boldsymbol{E}_{ij}$ at position $\boldsymbol{r}_i$ generated by the particle j at position $\boldsymbol{r}_j$ is given by

$$\boldsymbol{E}_{ij} = Q_j A(r_{ij}) \hat{\boldsymbol{r}}_{ij} \tag{3}$$

with the unit vector $\hat{\boldsymbol{r}}_{ij}$ given by

$$\hat{\boldsymbol{r}}_{ij} = \frac{\boldsymbol{r}_{ij}}{r_{ij}} = \frac{\boldsymbol{r}_j - \boldsymbol{r}_i}{|\boldsymbol{r}_j - \boldsymbol{r}_i|} \ . \tag{4}$$

Since we are not really interested in intricate dynamic behaviour the distance dependence of the field has been eliminated by setting $A(r) = 1$. The total field $\boldsymbol{E}_i$ at position i is obtained by vectorial addition:

$$\boldsymbol{E}_i = \sum_{j \neq i} Q_j \hat{\boldsymbol{r}}_{ij} \ . \tag{5}$$

The force $\boldsymbol{F}_i$ acting on particle i at position $\boldsymbol{r}_i$ is then given by

$$\boldsymbol{F}_i = Q_i \boldsymbol{E}_i \ . \tag{6}$$

As a consequence of this force the particle will move. For the equation of motion we define

$$\mu \dot{\boldsymbol{r}}_i = \boldsymbol{F}_i \ . \tag{7}$$

The acceleration term has been omitted because of computational convenience since, again, we are not really interested in detailed dynamics. Physically, it means that in our representation the particles move in a medium with high viscosity μ: the velocity of the particle is proportional to the force acting upon it.

The resulting displacement, finally, is obtained by simple numeric integration (Euler) using a time step δ:

$$\boldsymbol{r}_i(t + \delta) = \boldsymbol{r}_i(t) + (\delta/\mu)\boldsymbol{F}_i(t) \ . \tag{8}$$

This computational scheme is applied to all particles repeatedly as often as necessary to cover the duration of the multi-unit recording. The result is a collection of trajectories of the N particles in N-space.

Note that actually the movement is restricted to $N - 1$ dimensions since no external forces are involved. Furthermore, since all interactions are symmetrical, the center of mass of the system of particles is preserved.

2.3 Attraction and Repulsion

The gravitational rules given so far will necessarily lead to a collapsing universe. This is caused by the charges being definite non-negative [Eqs. (1–2)]: all particles attract and will eventually aggregate into the system's center of mass. Although it is certainly possible to infer functional proximity from studying the dynamics of evolving clusters, i.e., the time it takes the particles to aggregate (Wright 1977), we chose a slightly different approach. For our purpose we require particles which correspond to independently firing neurons to neither attract nor repel one another when studied over sufficiently long time. In this way only those particles will cluster which represent neurons that fire in a truly coherent fashion. The other ones will be subject to purely random influences; no systematic movement will occur: random walk with expected displacement equal to zero.

One way to obtain this behaviour is to impose an overall force on the particles which points outwards from the center of mass. This outward "drift" can be accomplished in several ways. In the present formulation of the gravitational representation we modify the charge rule [Eqs. (1–2)] to this effect: the original charge $Q(t)$ is replaced by shifted version $Q'(t)$, such that the time averaged value of $Q'(t)$ equals zero:

$$Q'(t) = Q(t) - \overline{Q(t)}. \tag{9}$$

The bar denotes taking the time average. A more general way to obtain zero mean charge would be to modify the impulse response $q(s)$ [Eq. (2)] to have no DC-component: the lowpass filter should be made bandpass to reject extremely low frequencies.

As a result of this modified charge rule the time averaged force for a system of two particles corresponding to two independently firing neurons, which is proportional to $\overline{Q_i'(t)Q_j'(t)}$, will be zero. When both neurons have fired recently, both charges Q' are positive: the particles will attract. When one neuron has fired recently and the other one remained silent, the signs of the charges differ: the particles will repel, the magnitude of the force being smaller than in the former case. Finally, when neither of the neurons fired recently, both charges will be negative: the particles will attract, the magnitude of the force dropping once more. These are precisely the properties needed to insure aggregation of those particles whose neurons either tend to fire in synchrony, or (more weakly) tend to be silent in synchrony. Particles that correspond to neurons which fire in an uncorrelated way will, on the average, exert no net force onto one another. Finally, particles whose neurons fire in anti-synchrony (antagonist type firing) will show a net divergent force.

2.4 Rate Effects

The original charge rule [Eqs. (1–2)] induces another peculiar property of the charge $Q(t)$. Due to the fact that for every particle the charge increases with an equal amount q_0 whenever the corresponding neuron fires, the propulsive influence of the charge would be proportional to the neuron's mean firing rate. Such a rate dependence may not be desirable. Therefore, in the original description of the gravitational representation (Gerstein et al. 1985), a "rate normalization" was applied: the increment of charge added at the time of each action potential was made proportional to the mean interspike interval for the neuron represented by that particular particle. In this way the mean value of charge $Q(t)$, considered over the whole recording (or a sufficiently long interval in the case of a sliding normalization) is the same for all particles; consequently the possibly undesirable effect of rate differences between neurons on the time averaged propulsive influence of the corresponding charges is canceled.

The more fundamental modification of the charge rule discussed earlier in connection with attraction and repulsion, i.e. going from $Q(t)$ to $Q'(t)$, causes a "rate normalization" in the above mentioned sense to become superfluous. Since the time averaged charge $Q'(t)$ by definition is zero, the overall effect of mean firing rate on propulsive "potential" no longer exists. More precisely: with equal increments in charge q_0 for every action potential clearly the magnitude of fluctuations in the charge $Q'(t)$ (the deviation from the expected value zero) will be proportional to the neuron's mean firing rate; however, the time avearage of the charge, and hence its global propulsive influence, is insensitive to it. As a consequence, when considering the force between any two particles it is not the rate of firing as such but truly the rate of "near-coincident" firing from both neurons which determines the net movement that will result. The only remaining effect of firing rate per se is on the "noisiness" of the trajectories, not on the systematic trends.

Not only is an additional "rate normalization" thus made superfluous, what is more, it even becomes counterproductive in the case of genuine correlated firing. The combination of the "rate normalization" and the transformation from Q to Q' [Eq. (9)] effectively amplifies the propulsive influence of "near-coincident" firings when one or more neurons with low mean firing rate are involved, while reducing it for neuron(s) with higher mean rate(s). This thus creates an undesired effect on the velocity of particle aggregation.

From the foregoing reasoning we conclude that, as far as the rate of coalescence is concerned, a charge rule without any specific "firing rate normalization" is probably the best choice for unbiased analysis of unknown data. This same conclusion was arrived at recently by a somewhat different reasoning (Gerstein and Aertsen 1985). It was pointed out there also, however, that the choice for a normalization rule must take into account additional factors, such as different selectivity for specific spike patterns. Results shown in the present paper were obtained while using a charge increment normalization as described by Gerstein et al. (1985).

3 Results

In this Section we show some results of applying the gravitational representation to simulated spike trains, generated by a simple neural network simulator (Gerstein et al. 1985, Aertsen and Gerstein 1985). More results can be found elsewhere (Gerstein et al. 1985, Gerstein and Aertsen 1985).

The aggregation process in the gravitational representation can be evaluated by studying the time evolution of the distance between pairs of particles. This should provide an indication about the functional connectivity of the corresponding neurons. Figure 4 shows the results for a simple neuronal circuit consisting of a pair of synaptically connected neurons together with a set of independently firing control neurons. The strength of the excitatory synaps was varied in the simulation, using values of 0.25, 0.50, and 0.99 (on a linear scale between 0 and 1) for different runs. Mean firing rates were in the order of 10 events per second. The figure shows, for each case, the time course of the distance between the points corresponding to pre- and postsynaptic neurons as well as those of an unrelated pair. The particles corresponding to unrelated neurons move about in a random fashion, their mutual distance remaining approximately constant. The particles corresponding to the connected pair of neurons, however, clearly show a systematic, albeit irregular, tendency

Fig. 4. The time to gravitational aggregation decreases as the strength of synaptic connection increases. Distance between selected pairs of points is shown as a function of step number (time) in the gravitational calculation. All pairs start at the distance of 100 arbitrary units; each time step corresponds to 2 ms of real time. The time constant for the charge decay was set at 10 ms, the distance moved per unit force was 3.5×10^{-5} units. The *three descending curves* show distances between particles that represent neurons connected with (*left to right*) decreasing excitatory synaptic strength of 0.99, 0.50 and 0.25, respectively. Distances between typcial particles representing independently firing neurons are at the *top* of the figure and remain essentially unchanged as the computation progresses. (From Gerstein et al. 1985)

to approach each other. The speed of aggregation appears to be monotonically related to the strength of the connection: the stronger the connection, the faster the coalescence. Note that coalescence was obtained already after about 40 spikes (4 s) in the case of strongest connectivity; the weaker interactions need approximately 7 and 10 s, respectively.

A more complex and interesting network of (simulated) neurons is shown in Fig. 5. The network consists of (1) two independently firing presynaptic neurons, (2) a set of four postsynaptic neurons for each presynaptic one, two neurons in each of these sets being driven by both presynaptic neurons, and (3) two independently firing control neurons. This example provides direct synaptic interaction, as well as two degrees of shared input. Firing rates of the individual neurons again were in the order of 10 spikes per second, connections were excitatory with a strength of 0.35, which puts them into the weak to moderate category encountered in physiological recordings. The time evolution of the clustering process is illustrated in Fig. 6, where, once again, pairwise distances are plotted as a function of time. The identities of the particles involved are indicated at the right hand side. Going from the top of Fig. 6 to the bottom, we observe the following: The distance between the particles 9 and 10 (unrelated control neurons) remains essentially constant throughout the entire run, which covers 16 s of "recording". The distances for the pairs (9,3) and (9,4) (one control neuron and in each case one postsynaptic neuron) initially decrease, later increase. The distances for the pairs (10,3) and (10,4) (the second control

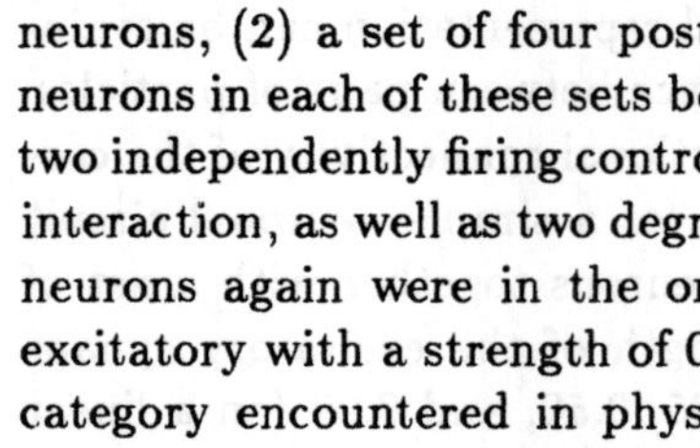

Fig. 5. A Venn diagram of more complicated neuronal connectivity that was simulated in a set of 10 spike trains. These data are analyzed in Figs. 6 and 7. Neurons *1* and *2* are drivers. Neurons *7* and *8* receive excitatory input from both drivers, while neurons *3,5* and *4,6* receive excitatory input only from one driver. Neurons *9* and *10* are independent control neurons. (From Gerstein et al. 1985)

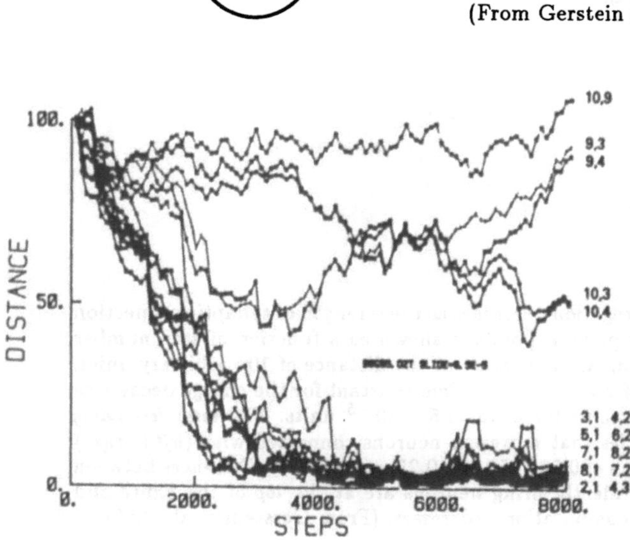

Fig. 6. Gravitational computation for the 10 spike train data set generated by the simulated neuronal circuit in Fig. 5. Ordinate: distance between particles representing selected neuron pairs; abscissa: time steps in the computation (2 ms step^{-1}). Further details in text. (From Gerstein et al. 1985)

neuron and one postsynaptic neuron) slowly decrease through the calculation. Finally, the pair distances for various transsynaptic pairs, for the two drivers (1,2) and for an independent postsynaptic pair (4,3) all show similar rapid coalescence. Full aggregation is obtained after some 8 s. Analysis of the same data, but excluding the spike sequences from the driver neurons 1 and 2 leads to essentially the same result (not shown here): curves look very similar, the main difference being that aggregation is somewhat slower. In the latter case full aggregation is reached after about 12 s (Gerstein et al. 1985). This shows that also in case the multi-unit recording fails to include the presynaptic driver neurons the interacting neurons can clearly be distinguished from the control neurons.

All synaptic connections in the spike generating network (Fig. 5) have been set to be equally strong. Thus each driver attracts equally (and is attracted by) each of its postsynaptic partners. Indirectly this will bring all of the postsynaptic partners of one driver neuron together in one small cluster, quite independent from the degree of direct interaction between the postsynaptic neurons. For similar reasons particles 7 and 8, which both are connected to both drivers 1 and 2, through the symmetry of the gravitational interaction will act as attractors for the clusters around the drivers. This eventually leads to clustering of all particles 1 to 8, although, for instance, the pairs (1,2) (the drivers) and (3,4) (unrelated postsynaptic neurons) represent neurons which, as such, are not in any way functionally related. Apparently the analysis of pair distances can identify a subset of interacting neurons; in itself it is insufficient to infer the circuit or the strength of connections in it.

The investigation of pair distances is a rather primitive tool to study the information that is contained in an ensemble of trajectories in N-space. More insight may be gained by adding one dimension in the representation of results: visualization of the trajectories by projection on an appropriately chosen plane. Figure 7 shows the results of a projection where information about the circuitry was used to define an adequate projection plane within the N-space: the plane determined by the instantaneous positions of the particles 1 and 2 (the drivers) and the center of mass of the particles 9 and 10 (control neurons). Note that, since these points are travelling through N-space, also the projection plane itself will be moving. Figure 7a–f shows "snapshots" of the projected positions of the 10 particles at regular intervals of 1.6 s. From the definition of the projection plane it follows that the particles 1 and 2 will move inwards in a straight line. The particles 9 and 10 (control neurons) can be observed to move rather randomly in the lower half of the pictures. All other particles (the driven ones) start at a point near the middle of the screen, and are rapidly dragged towards the aggregating drivers 1 and 2. The sensitivity of the gravitational representation is quite dramatically demonstrated in Fig. 7c: already after 4.8 s (i.e., after a mere 50 spikes from each neuron) the projection shows a configuration of particles that is reminiscent of the Venn diagram describing the network in Fig. 5. Each driver has collected its own exclusively driven particles, while the jointly driven particles 7 and 8 form a "bridge" between the

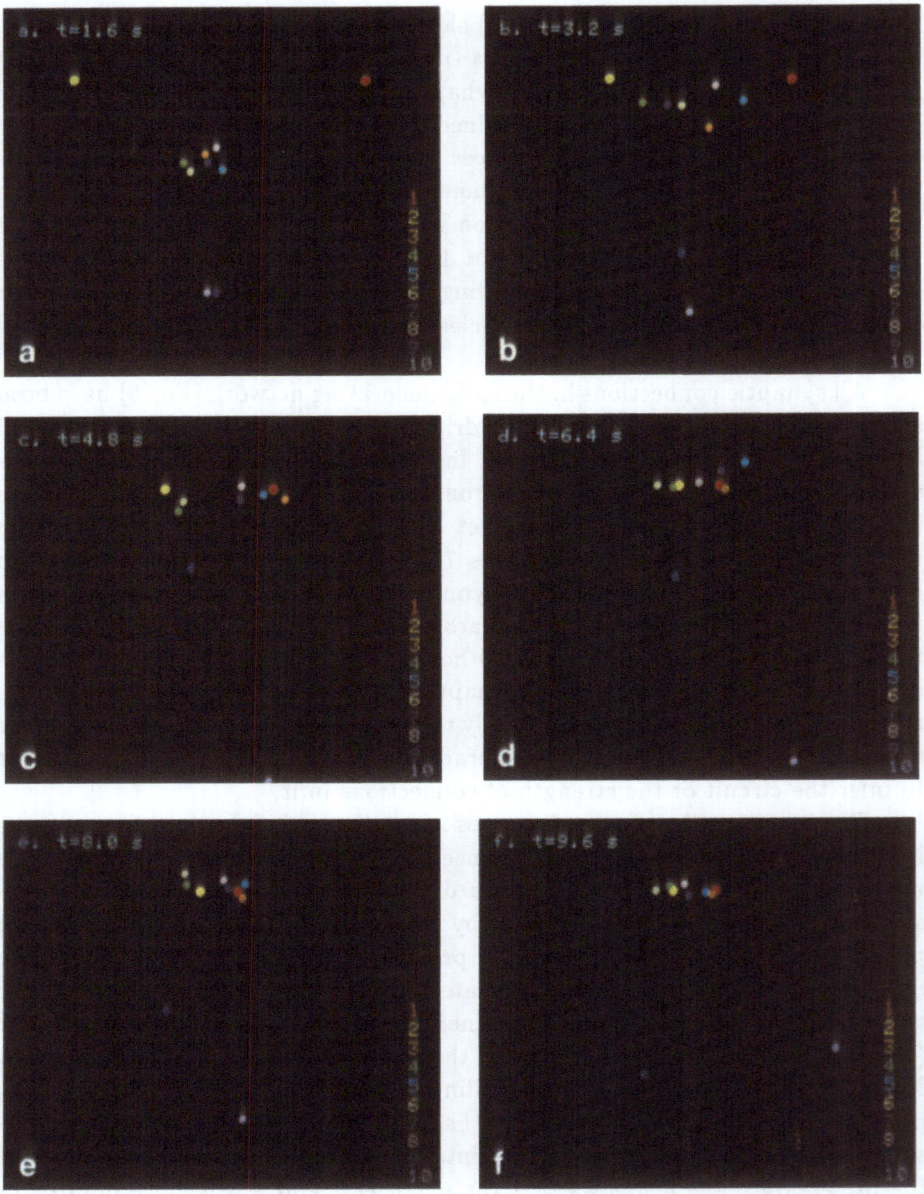

Fig. 7. Visualization of gravitational clustering by two-dimensional projection. Spike trains were obtained from the simulated network of 10 neurons, shown in Fig. 5. The projection of the *10 points* (representing the 10 neurons) from the 10-space onto an appropriate plane is shown in the form of "snapshots", taken at different moments in time with regular intervals of 1.6 s of "real" time. Particles are identified by their colour and the corresponding neuron number at the *right hand side* of each figure. Already after 4.8 s (i.e., after roughly 50 spikes from each neuron) the projection in Fig. 7c shows the Venn-diagram of the original network (cf. Fig. 5). As time progresses the configuration continues to aggregate, however, maintaining the basic character of the Venn-diagram (Figs. 7d–f). Further details in text

primary clusters. As time moves on, the configuration "boils" and continues to aggregate (Fig. 7d–f) until one single cluster remains, which still shows the characteristic structure of the Venn diagram, already clear in Fig. 7c.

A very good impression of the dynamics of the aggregation process can be obtained by combining a sequence of such snapshots into an animation movie. In such a movie especially the role of the "attractors" and the further evolution of the separate clusters becomes quite manifest. A more elaborate discussion on the possibilities of visualization by projection, also in the absence of prior information about the underlying network (used here to define the projection plane) can be found elsewhere (Gerstein and Aertsen 1985).

4 Discussion

4.1 Gravitational Clustering

We have described a new conceptual approach to representing and analysing spike train data from a multi-unit recording experiment. As shown by the examples the method is able to analyse interrelationships within a group of observed neurons without having to decompose the group into pairs as is usually done. The strength of interaction between neurons, reflected in the degree of firing synchrony, is manifested in the velocity of aggregation of the corresponding particles. The aggregation process can be studied by investigating the time dependence of interpair distances or, more visually directed, by projecting the trajectories onto an appropriately chosen plane in N-space.

A number of interesting issues concerning this representation could not be addressed here. These include, amongst others, the detection of inhibitory interactions and a modification of the gravitational clustering such that it allows "causal" inference: representation of the network in terms of a directed graph, rather than just identifying which neurons show associated firing behaviour. These and other issues are discussed at greater depth in the original papers (Gerstein et al. 1985, Gerstein and Aertsen 1985).

4.2 Sensitivity

The sensitivity of the method is extraordinary as becomes apparent from Fig. 7c, where a mere 50 spikes from each contributing neuron suffices to produce aggregation into a particle configuration which shows the essential characteristics of the circuit's Venn diagram. This sensitivity transcends at least by an order of magnitude that of usual measures like the Joint PST scatter diagram (Gerstein and Perkel 1972) or the three neuron "snowflake" (Perkel et al. 1975) which require hundreds to thousands of spikes from each neuron to produce clear pictures, moreover with the constraint that only two or three neurons are analyzed simultaneously.

The basic reason for this improvement is related to the different effects of time integration in these methods. In the gravitational representation the net distance traveled by a particle is proportional to the time integral of the total of "interactive force components" Q_iQ_j acting upon it. As a result of the definition of the charge function, especially the modification of the type as given in Eq. (9), the integrated effect of "noisy" spikes on the average cancels to a zero displacement; only the "near-coincident" firings result in a net effect on the particle positions. This "differential" behaviour of the gravitational representation is in contrast with usual correlation type methods as mentioned above, where time integration results in a steadily increasing noisy background in the scatter diagrams. The systematic effects of correlated firing usually are hardly discernible, unless many spikes have been processed to reduce the variance in the background to a comfortable enough level.

4.3 Type of Interactions

It should be observed that the force rule defined in Eqs. (3–6) only takes into account pair interactions, i.e., interactions of the type Q_iQ_j. This implies that the forces, and thus the trajectories of the particles, are determined by pair correlations of the corresponding spike trains, i.e., correlations of the type z_iz_j. In other words, specific higher order correlations in the neural activity, e.g., of the sort $z_iz_jz_k$, which go beyond combinations of participating doublets, are not considered in this representation.

The vectorial addition of force components from different sources (Eq. 5), although combining the influence of different doublets, by its linear nature does not address the possible presence of purely higher order interactions. Analogously to the correlation approach to nonlinear systems (.e.g., Marmarelis and Marmarelis 1978, Eggermont et al. 1983b) one might say that the gravity representation specifically aims at highlighting that component in the neural interaction which can be described by linear spatio-temporal integration (spatial = summation over different neurons; temporal = (leaky) integration over past spike activity from observed neurons (e.g., Johannesma and Van den Boogaard, 1985)). As a consequence specific nonlinear (e.g., Poggio and Torre, 1981) or composed (e.g., Shepherd 1974) synapses may go undetected.

The principal advantage of the gravitational representation as compared to normal cross correlation of pairs as applied extensively in the analysis of multi-unit data (e.g., Gerstein 1970), is that all possible pair interactions are analyzed at the same time. This is precisely what is accomplished by the multi-dimensionality of the present approach.

Another interesting observation to be made, is that the gravitational equations described above bear a close resemblance to the formalism describing plasticity of synaptic coupling between neurons, "Hebb's rule" (Hebb 1949, Palm 1982). In the latter context the principal drive for "learning", i.e., changing the synaptic weight, is supposed to reside in pair interactions of participating neurons (coherent pre- and postsynaptic firing). Like the gravitational representation, Hebb's rule does not take into account triple or higher order interactions. One might say that gravitational clustering considers neuronal interaction in a Hebbian universe.

This analogy might be pushed even further by imagining a somewhat different realization of the gravitational representation. Instead of working with particles in an N-space one might represent the N neurons under observation as a network of N nodes with some initially specified connectivity matrix (e.g., uniform or random). Applying the formalism described in the present paper, with only slight modifications, to gradually "update" the connectivity matrix, this network could be made to "learn" by changing its "synapses", such that the final state would more or less mimic the circuit of neurons that generated the measured spike sequences in the first place.

4.4 Stimulus-Dependent Neural Interaction

Until now it was implicitly assumed that the pattern of neural interactions remains fixed throughout the recording. An interesting question arises when this pattern, in fact, does change, for instance induced by changing stimulus conditions (see the example given in the Introduction). This should be reflected in a change in the aggregation process in the gravitational representation. Clusters may form and, later in the calculation, may "evaporate" to build new configurations, more or less mimicking the way the corresponding neurons are involved in a dynamic succession of different assemblies, "ignited" by changes in the external world (Braitenberg 1977). The dynamics of clustering thus represent aspects of structure (anatomy) as well as process (physiology). A third aspect, the function of neuronal interaction, can only be understood once the relation with the outside world, i.e., the stimulus ensemble giving rise to the observed phenomena, is explicitly taken into account. The analysis of multi-unit activity by gravitational-like methods emphasizes the "syntactical" aspects of neural activity. Stimulus-response correlation focuses on the "semantical" aspects. It is in the interrelation of both approaches that meaningful models of sensory information processing may be hoped to emerge.

Acknowledgements. This work was supported through grants from NIH (GLG NS05606), the System Development Foundation (GLG SDF 0013), the Fulbright Program (AA travel grant) and ZWO (AA,PJ). The help of Volker Staiger in obtaining adequate copies of the Figures is gratefully acknowledged.

References

Aertsen AMHJ, Gerstein GL (1985) Evaluation of neuronal connectivity: sensitivity of cross-correlation. Brain Res 340:341–354

Bloom M, Gerstein G (1984) Multineuron analysis shows spatial tuning not found in single unit responses. Soc Neurosci Abstr 10:245

Boogaard HFP van den, Hesselmans GHFM, Johannesma PIM (1985a) Transformation of point processes, correlation functions and system identification. Submitted

Boogaard HFP van den (1985b) Neural interaction equations: Transformation of point processes. Ph.D. Thesis, University of Nijmegen, Nijmegen, The Netherlands
Braitenberg V (1977) Cell assemblies in the cerebral cortex. In: Heim R, Palm G (eds) Theoretical approaches to complex systems. Lecture notes in biomathematics, vol 21. Springer, Berlin Heidelberg New York, pp 171–188
Craik K (1943) The nature of explanation. Cambridge Univ Press, Cambridge
Eggermont JJ, Epping WJM, Aertsen AMHJ (1983a) Stimulus dependent neural correlations in the auditory midbrain of the grassfog (Rana temporaria L.). Biol Cybernet 47:103–117
Eggermont JJ, Johannesma PIM, Aertsen AMHJ (1983b) Reverse-correlation methods in auditory research. Q Rev Biophys 16:341–414
Espinosa I, Gerstein G (1984) Correlation among spike trains in cat's auditory cortex during presentation of three-tone sequences. Soc Neurosci Abstr 10:245
Evans EF (1974) Neural processes for the detection of acoustic patterns and for sound localization. In: Schmidt FO, Worden FG (eds) The neurosciences; 3rd Study Program, MIT Press, Cambridge, pp 131–145
Frostig RD, Gottlieb Y, Vaadia E, Abeles M (1983) The effects of stimuli on the activity and functional connectivity of local neuronal groups in the cat auditory cortex. Brain Res 272:211–221
Fukunaga K (1972) Introduction to statistical pattern recognition. Academic Press, London New York
Gerstein GL (1970) Functional association of neurons: detection and interpretation. In: Schmitt FO (ed) The neurosciences: 2nd Study Program. Rockefeller Univ Press, New York, pp 648–661
Gerstein GL, Perkel DH (1972) Mutual temporal relationships among neuronal spike trains. Biophys J 12:453–473
Gerstein GL, Bloom MJ, Espinosa IE, Evanczuk S, Turner MR (1983) Design of a laboratory for multineuron studies. IEEE Trans Syst Man Cybern SMC-13:668–676
Gerstein GL, Perkel DH, Dayhoff JE (1985) Cooperative firing activity in simultaneously recorded populations of neurons: detection and measurement. J Neurosci 5:881–889
Gerstein GL, Aertsen AMHJ (1985) Representation of cooperative firing activity among simultaneously recorded neurons. J Neurophysiol (in press)
Grinvald A (1984) Real-time optical imaging of neuronal activity. Trends Neurosci 7:143–150
Hebb DO (1949) The organization of behavior. A neuropsychological theory. Wiley, New York; Chapman and Hall, London
Johannesma PIM, Eggermont JJ (1983) Receptive fields of auditory neurons in the frog's midbrain as functional elements for acoustic communication. In: Ewert JP, Capranica RR, Ingle DJ (eds) Advances in vertebrate neuroethology. Plenum Press, New York London, pp 901–910
Johannesma PIM, Boogaard HFP van den (1985) Stochastic formulation of neural interaction. Acta Math Appl 4:201–224
Marmarelis PZ, Marmarelis VZ (1978) Analysis of physiological systems. The white-noise approach. Plenum Press, New York London
McCulloch WS (1965) Embodiments of mind. MIT Press, Cambridge
Palm G (1982) Neural assemblies, an alternative approach to artificial intelligence. Studies of brain function, vol VII. Springer, Berlin Heidelberg New York
Perkel DH, Gerstein GL, Moore GP (1967) Neuronal spike trains and stochastic point processes. II. Simultaneous spike trains. Biophys J 7:419–440
Perkel DH, Gerstein GL, Smith MS, Tatton WG (1975) Nerve-impulse patterns: a quantitative display technique for three neurons. Brain Res 100:271–296
Poggio T, Torre V (1981) A theory of synaptic interaction. In: Reichardt W, Poggio T (eds) Theoretical approaches in neurobiology. MIT Press, Cambridge, pp 28–38
Shepherd GM (1974) The synaptic organization of the brain. Oxford Univ Press, New York
Wright WE (1977) Gravitational clustering. Pattern Recognition 9:151–166

From Synchrony to Harmony: Ideas on the Function of Neural Assemblies and on the Interpretation of Neural Synchrony

P. JOHANNESMA[1], A. AERTSEN[2], H. VAN DEN BOOGAARD[1], J. EGGERMONT[1], and W. EPPING[1]

1 Introduction

Point of departure are experimental data acquired by simultaneous recording of the activity of a number (2–16) of individual neurons during presentation of a sensory stimulus. The area under investigation is the auditory midbrain (Torus semicircularis) of the immobilized grassfrog (*Rana temporaria L.*). The sensory stimuli are both artificial (noise, tones and clicks) and natural sounds (vocalizations and environmental sounds). The goal of investigation is an insight into the neural representation of the sensory environment.

In a first approach the multi-unit recording is regarded as a set of separate recordings of single neurons. Second order correlation of auditory stimulus and neural events leads to the functional description of each neuron in terms of its spectro-temporal sensitivity (STS) in relation to a given stimulus ensemble. Making use of nonlinear system theory (i.e. Wiener functions and Volterra kernels) it turns out that for a fraction of the neurons the STS can be normalized with respect to the stimulus ensemble (SE). In this case a stimulus-invariant spectro-temporal receptive field (STRF) is defined and measured for the neuron (Aertsen and Johannesma 1981a,b, Eggermont et al. 1983a). If the SE is considered as a distribution of signals in acoustic space then the receptive field forms the subspace of this acoustic space where the neuron is most sensitive; its "focal zone" (Scheich 1977, Johannesma and Aertsen 1982). When, to a good approximation, a stimulus-invariant STRF does exist the focal zone of the neuron in acoustic space forms an inherent property of that neuron independent of sensory context. For many neurons, however, in the frog's auditory midbrain the STS for different stimulus-ensembles cannot be normalized into a unique STRF. This implies that, at least using the known methods of system theory, the concept of single unit receptive field (SURF) loses its well-defined and unique features as a functional characteristic of the sensitivity of a neuron for sensory stimuli (Johannesma and Eggermont 1983).

[1] Department of Medical Physics and Biophysics, University of Nijmegen, Nijmegen, The Netherlands
[2] Max Planck Institut für Biologische Kybernetik, Spemannstraße 38, 7400-Tübingen, FRG

Brain Theory
Edited by G. Palm and A. Aertsen

In a second approach the multi-unit recording is considered as the observation of dynamic patterns of activity from a set of neurons (Gerstein 1970, Gerstein and Michalski 1981). As such it may be the representation of the activity of a neural assembly (Hebb 1949). The interest now focuses on analysis and interpretation of neural synchrony, synchrony being defined as the spatio-temporal form of the neural activity pattern. To what extent are the events of different neurons correlated? Only as far as can be explained by the structure of the stimulus in combination with the single unit receptive fields or is there an additional correlation caused by neural interaction? This aspect of spatio-temporal correlation (coherence) is thought to be generated by a population of neurons interacting cooperatively (assembly) under the given environmental situation (Braitenberg 1977, Palm 1982). Recently investigations have been made of the characteristics of synchrony of firing in multi-unit recordings from the torus semicircularis of the frog, again under a variety of stimulus conditions. It appeared that neural correlation, i.e. after application of a stimulus normalization procedure, essentially based on an additive model, again may be stimulus-dependent. In a number of cases the neural correlation turns out to be stimulus-variant, possibly context-sensitive (Eggermont et al. 1983b, Epping et al. 1984).

One way to put these results together would be to attach "labels" to the SURF and the neural correlation, the label specifying the particular stimulus context in which the neuron characteristic applies. This "solution", however, from the point of view of modeling the central nervous system is far from attractive. As an alternative we propose the hypothesis that the stimulus-invariance of both the SURF and the neural correlation are intrinsically interconnected. In fact, they both are manifestations of one underlying mechanism: the neurons being elements in an interacting population, possibly giving rise to neural assemblies. From this point of view the activity of the neuron under experimental observation is, apart from the direct influence of stimulus and local connectivity at least partially influenced by the activity pattern in the (non-observed) remainder of the network, this remainder in turn is (at least partially) influenced both reciprocal and by direct stimulus influence. The net effect would be an apparent stimulus control of various neural characteristics like SURF and neural correlation, without, however, the necessity to assign this overall stimulus effect to a specific neuron, connection or local circuit. A strategic consequence of this hypothesis would be that for a fruitful approach of the question of stimulus-variance of the SURF and neural correlation both aspects should be addressed in an integrated manner: stimulus-event correlation on its own, as well as event-event correlation on its own will only lead to a multitude of images, each one with its associated context of "validity". It is in the interrelation of stimulus-event relations and neural interaction that meaningful models, i.e. with a functional meaning associated to them, may hopefully emerge.

The composition of this exposition is as follows. In Sect. 2 experimental data from multi-unit recordings are presented as well as some transformations of these data. General ideas from system theory are given in Sect. 3 and the difference between reactive and creative systems is introduced. In Sect. 4 equations for neural interaction and synaptic plasticity are summarized. Some concepts and speculations with respect to perception, brain and behaviour are presented in Sect. 5. Finally in Sect. 6 an experimental procedure is proposed for the investigation of neural assemblies in relation to perception.

The main theme of this presentation is a discrimination of two ways of analysis and interpretation of multi-unit recordings. On one hand the "syntactical" approach as introduced by Gerstein and coworkers relating structure and process of the nervous system. On the other hand the "semantical" method as described before by Johannesma (1981) and Johannesma and Aertsen (1982) trying to connect process and function of the nervous system. The contribution of Aertsen et al., in this Vol. and this contribution are intended as integrated complementary views on the treatment of multi-unit recordings.

2 Registration and Representation of Neural Activity Patterns

Experimental data as recorded in the auditory midbrain of the immobilized grassfrog are yet far from a complete observation of activity patterns of a neural population. At this moment we can present the results of simultaneous recordings of up to four neurons under different stimulus conditions. These data are insufficient for investigation of the actual role of neural assemblies but, provided neural assemblies do occur in this region of the brain of the frog, they may well form the base for a delineation of the questions which have to be treated if large scale multi-unit recordings become available. For a description of experimental methods see Eggermont et al. (1983c) and for data analysis and results Epping et al. (1984).

Given the measurement of the simultaneous activity of four neurons during presentation of a sequence of tones with different frequencies, the first goal is the presentation of these data in such a way that the identity of the neurons is preserved. This is realized by using colour as the code for identification of each neuron. The resulting display, the "neurochrome", gives in an integrated way the activity of the neurons (colour-dots) as function of time (horizontal) and the number of the tonal stimulus (vertical). The results for a quadruple recording are depicted in Fig. 1. In order to prevent sequential effects of the tonal series the frequencies where presented in a pseudo-random order. Rearrangement of the responses of the neuron with respect to increasing/frequency results in the ordered neurochromes shown in Fig. 2. The results show clearly that the time structure of the response to tones depends on the frequency of the tone and that this dependence is different for different neurons. It should be noted that the four neurons were recorded on one micro-electrode, i.e. they were in close proximity.

The neurochrome contains the full information concerning the stimulus-associated neural activity patterns as far as observed through the multi-unit

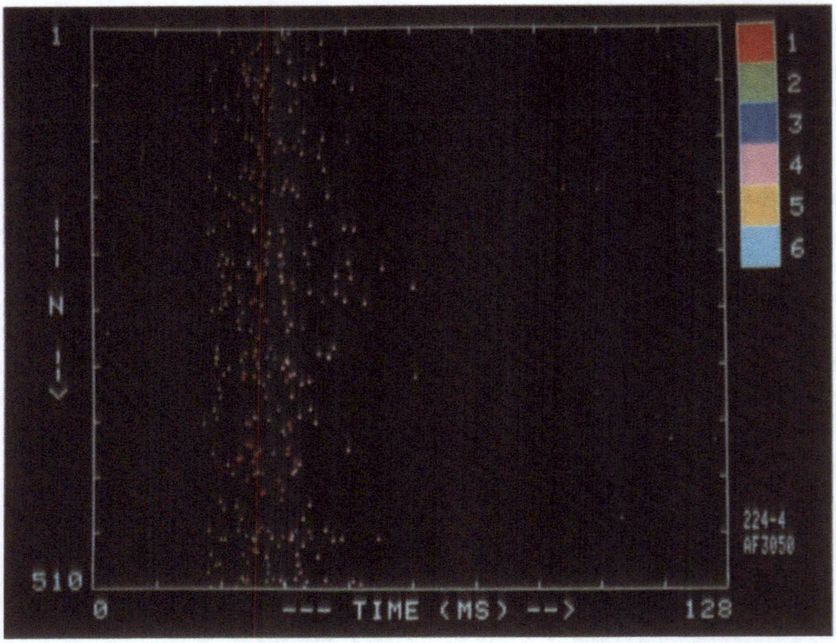

Fig. 1. Chromatic representation of the simultaneous measurement of action potentials of four neurons in the Torus Semicircularis of the grassfrog during tonal stimulation. *Horizontal axis* represents time, *vertical axis* the sequential index of the tones which were presented in random order

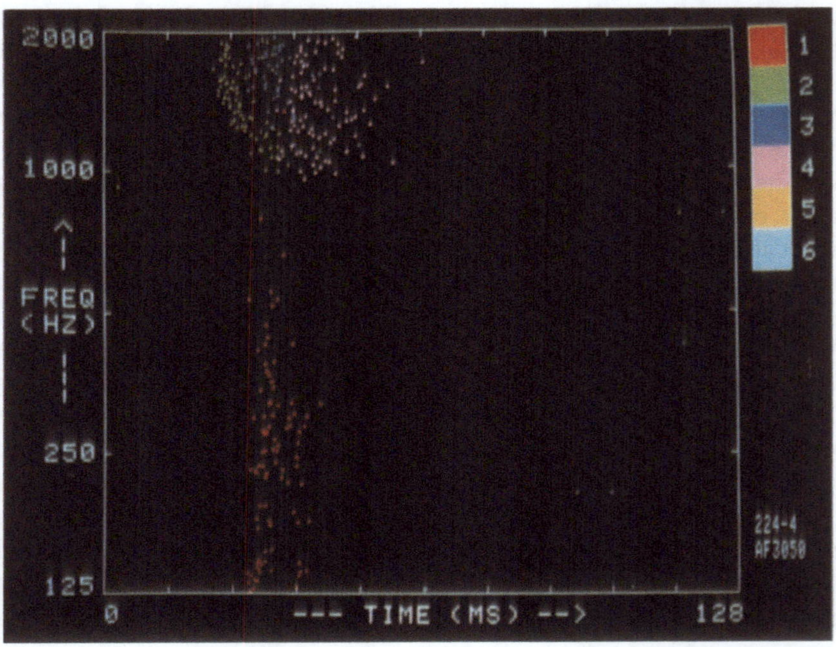

Fig. 2. Chromatic representation of the simultaneous measurement of action potentials of four neurons in the Torus Semicircularis of the grassfrog during tonal stimulation. *Horizontal axis* represents time, *vertical axis* the frequency of the tone. Note different response areas of different neurons

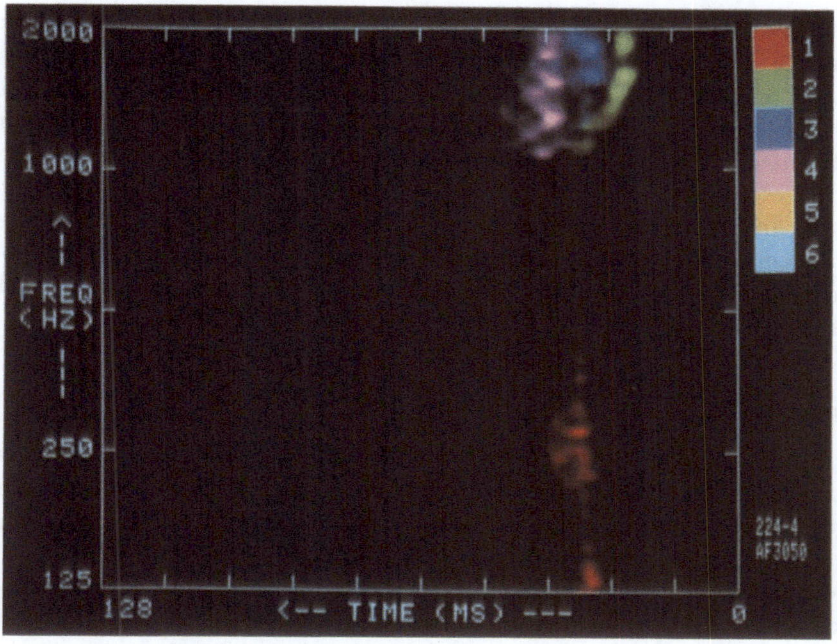

Fig. 3. Chromatic representation of spectro-temporal sensitivity of four neurons in the Torus Semicircularis of the grassfrog recorded simultaneously during tonal stimulation

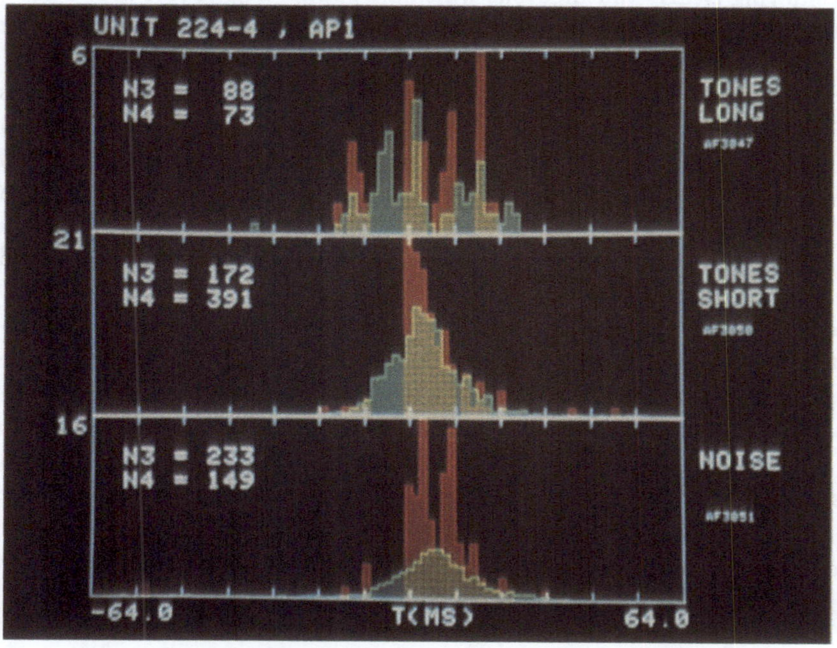

Fig. 4. Simultaneous and nonsimultaneous cross-coincidence histograms of a pair of neurons in the Torus Semicircularis of the grassfrog under different stimular conditions

recording (SAMUR). For the evaluation of the SAMUR two ways are open: a "syntactical" and a "semantical" analysis. The syntactical one concentrates upon the form of the activity patterns by means of analysis of the correlation of activity of different neurons: event-event correlation. The semantical approach is directed toward the meaning of the neural activity through a study of the association of sensory stimulus and neural action potentials: stimulus-event correlation. In both points of view action potentials as generated by a single neuron are considered as events: irreducible elements for representation of sensory stimuli. Form and size of an action potential are irrelevant, only when and where it occurs is significant.

2.1 Semantic Analysis of Neural Activity

For a given neuron the action potentials indicate the selection of a stimulus out of the ensemble of stimuli presented to the animal. This selection can be investigated by constructing the pre-event stimulus ensemble (PESE) which forms a subset of the stimulus ensemble (SE).

The characteristics of the PESE in relation to the SE represent the stimulus-selective function of the neuron (Johannesma 1972, 1980, 1981, Johannesma and Aertsen 1982, Eggermont et al. 1983c). A plausible way to determine these characteristics is to perform a spectro-temporal averaging of the PESE. The result of this procedure is the spectro-temporal sensitivity (STS) of the neuron (Aertsen and Johannesma 1981a). In Fig. 3 the STS is given for different neurons showing clearly that neurons which are spatial neighbours in the auditory midbrain do not have to be spectro-temporal neighbours in acoustic space: topography does not appear to be preserved.

For the definition of a single unit receptive field (SURF) the characteristics of the PESE should be normalized with respect to the SE and then result in a stimulus-invariant spectro-temporal sensitivity. The theoretical aspects and mathematical equations of this procedure are given in Aertsen and Johannesma (1981a) and comparative results for tonal and natural stimuli are presented in Aertsen and Johannesma (1981b) and Johannesma and Eggermont (1983) and Aertsen et al. (this Vol.).

The conclusion from the experimental data is that for a considerable fraction of the neurons in the auditory midbrain of the immoblized grassfrog the normalized spectro-temporal sensitivity does depend on the choice of the stimulus ensemble. A stimulus invariant single unit receptive field does not anymore exist in this central region of the frog's brain. A plausible explanation for this result is the presence of not only afferent connections to these neurons: the neuron forms part of a system where there is no longer an exclusive forward stream of information, lateral and recurrent interaction between neurons may become influential. We will return to this point in Sect. 4.

2.2 Syntactic Analysis of Neural Activity

In this approach the action potentials are again considered as events: irreducible elements for communication among neurons. A neural activity pattern is considered as a parallel series of point processes or "marked point process" (Snyder 1975). The focus of interest is now the internal structure of this marked point process. In a qualitative way this amounts to a study of coincidences of events of different neurons shown in Fig. 1. For a systematic approach we may search for the structure of each neural spike train (singlet), that of each pair of neurons (doublet), that of three neurons at a time (triplet) etc.

Some examples of pair correlations are given in the coincidence histograms in Fig. 4. In order to get an impression of the interaction giving rise to this correlation we compare simultaneous and non-simultaneous cross coincidence histograms. As can be seen in Fig. 4 the form of the cross-coincidence does depend on the stimulus-ensemble which has been used. For certain classes of models of the neural interaction, essentially limiting this interaction to a non-reciprocal form, it appears possible to compute expressions for the normalization of the simultaneous coincidence by the non-simultaneous one (van den Boogaard et al. 1985). However both experimental data as recorded in frog's auditory midbrain (Eggermont et al. 1983c) and the theoretical considerations exposed in Sect. 4 indicate that in general the pair correlation will be stimulus-dependent.

A further problem for the systematic computation of multiple correlation is the fast increase of the number of correlations of multiplets of order k as function of the number N of neurons which have been observed simultaneously. As an illustration Table 1 gives these numbers for $N = 1 - 8$. Even if the computer is able to do all necessary calculations within an acceptable time, the number of resulting graphs or displays will exceed the cognitive and associative capacity of the observer. Moreover coincidences or doublets can be shown in a histogram, triplets in an event-display (snowflake) but for quadruplets and higher multiplets no form of display is known. This further limits the acquisition of insight by a human observer.

An alternative evaluation of the form of multi-unit activity patterns has been proposed by Gerstein and coworkers: "gravitational clustering". The grav-

Table 1. Number of correlation functions of order k for a group N neurons

$\binom{N}{k}$	$k = 1$	2	3	4	5	6	7	8
$N = 1$	1							
2	2	1						
3	3	3	1					
4	4	6	4	1				
5	5	10	10	5	1			
6	6	15	20	15	6	1		
7	7	21	35	35	21	7	1	
8	8	28	56	70	56	28	8	1

itational clustering method influences location of the neurons in a hypothetical space according to their synchrony of firing. See the contribution of Aertsen et al. in this Vol. for a detailed description of concepts, methods, results and for references. However, also with this approach some problems do exist. Up till now only pair correlations have been taken into account in the clustering algorithm; how to include multiple correlations? The dimensionality of the hypothetical neural space is apriori equal to N; even if by the clustering the dimensionality reduces, it may well remain larger than three, leading to serious representational difficulties.

2.3 Conclusions

The conclusions from the experimental point of view for central regions in sensory parts of the brain are:

1. Relation of sensory stimulus and neural events depends on choice of stimulus-ensemble: a stimulus-invariant single unit receptive field in many cases does not exist.
2. Correlation of activity of different neurons depends on stimulus ensemble.
3. No general method exists for the analysis and interpretation of stimulus associated multi-unit recording.

3 System Theory

In order to treat the stimulus-associated multi-unit recording (SAMUR) some general ideas on typology of systems are exposed. The brain may be considered as a multi-input/multi-output system composed of nonlinear stochastic elements. Sensory stimuli, e.g. sound, form the input, behavioural activity is the output. Two types of global feedback are present: internal and external. Internal receptors monitor the behavioural activity (propriocepsis) and behavioural acts change the external environment and/or the position of the animal in this environment both leading to changes in the sensory stimulus. A sketch is given in Fig. 5.

For a more detailed characterization of the brain as a system it is relevant to make a distinction into "*reactive*" and "*creative*" systems. We assume that the system consists of N elements each characterized by a state variable u_n. Neglecting for the moment the stochastic aspects, the system may be represented by a set of N-coupled nonlinear differential equations. Now the relevant point is if by a well-chosen labeling of the elements the system may be drawn in such a way that there exist only forward influences. If this possibility does exist the system is defined as "reactive". The mathematical description then becomes

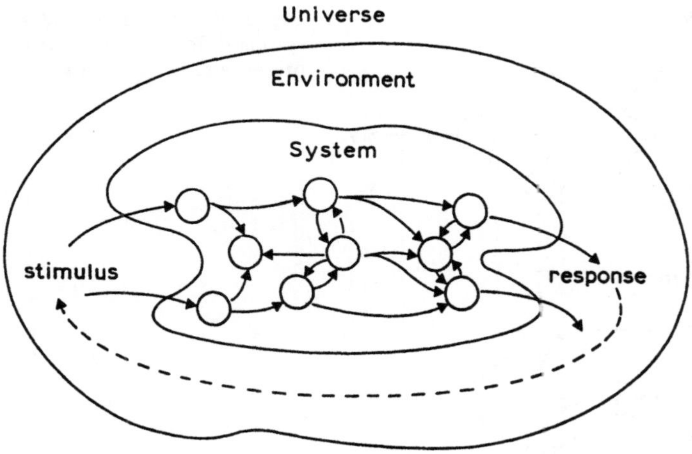

Fig. 5. Sketch of a system as a set of elements and relations in interaction with its environment and embedded in a universe

$$
\begin{aligned}
\frac{d}{dt}u_n &= f_n(x; u_1, ..., u_n) \quad n = 1, N \\
y &= g(x; u_1, ..., u_N)
\end{aligned}
\tag{3.1}
$$

where x = input, y = output, and u_n = state of element n.

If such a unidirectional description does not apply because of the presence of closed loops caused by mutual interaction and/or recurrent connections, then the system is defined as "creative". The mathematical description cannot be reduced to the form given in Eq. (3.1) but reads instead

$$
\begin{aligned}
\frac{d}{dt}u_n &= f_n(x; u_1, ..., u_N) \quad n = 1, N \\
y &= g(x; u_1, ..., u_N)
\end{aligned}
\tag{3.2}
$$

Illustrations of the structure of a reactive and a creative system are given in Fig. 6. The difference between Eq. (3.1) and Eq. (3.2) is solely in the index n or N for the differential equations; however, this has far reaching consequences. Equation (3.1) can be written in a more explicit form

$$\frac{d}{dt}u_1 = f_1(x; u_1) \tag{3.1a}$$

$$\frac{d}{dt}u_2 = f_2(x; u_1, u_2) \tag{3.1b}$$

$$\frac{d}{dt}u_3 = f_3(x; u_1, u_2, u_3) \ . \tag{3.1c}$$

Equation (3.1a) is a first-order non-homogeneous nonlinear differential equation. Given the forcing function $x(t)$ and the initial value of $u_1(t)$ then under some restrictions for the function f_1, e.g.

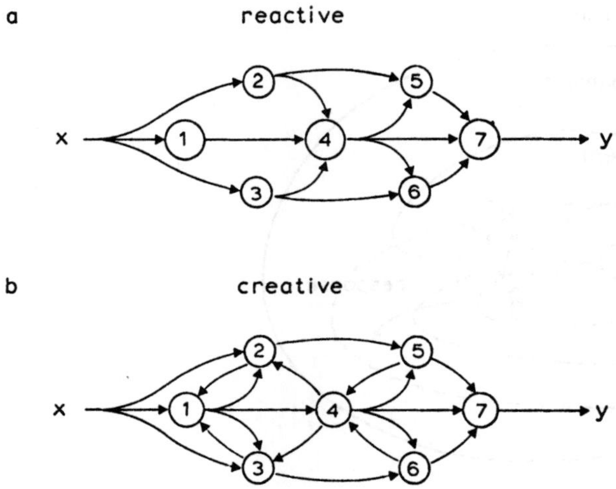

Fig. 6. Reactive and creative system differing in absence or presence of closed loops

$$f_1(x; u_1) = a(x) + b(u_1)$$

or

$$f_1/x; u_1) = c(x) \times d(u_1)$$

with again some restrictions on $b(u_1)$ and $d(u_1)$ Eq. (3.1a) can be solved in the form

$$u_1(t) = F_1[x(s), s < t] := F_1[x] \ . \tag{3.3a}$$

Substitution of Eq. (3.3a) into Eq. (3.1b) leads to

$$\frac{d}{dt}u_2 = f_2(x; F_1[x], u_2) := g_2(H_2[x]; u_2) \ .$$

In an analogue way and under analogue conditions this equation can be solved with the result

$$u_2(t) = F_2[x] \ . \tag{3.3b}$$

This approach can successively be applied to the complete set of Eqs. (3.1). However, because of its recursive form, such a sequential solution procedure cannot be applied to the set of Eqs. (3.2).

The general solution for the reactive system characterized by Eq. (3.1) has the form

$$u_n(t) = F_n[x(s), s < t; u_m(0), m \leq n] \tag{3.3}$$

indicating that the state of element n depends on the previous input and on the initial state of element n and of the preceding elements. The form of $b(u)$ or $d(u)$ determines if the influence of the initial state decays as time progresses. If this is the case, then Eq. (3.1) can be rewritten in an integral form such that each state variable u_n as well as the output y can be expressed as a sum and approximated by a finite sum of weighted integrals of power functions of the input x: the Volterra expansion.

$$u_n(t) = \sum_{m=0}^{M(n)} V_{nm}[x;t] \tag{3.4}$$

where

$$V_{nm}(t) = \int ds_1 \int ... \int ds_m v_{nm}(s_1, ..., s_n) \prod_{l=1}^{m} x(t - s_l)$$

and an analogous expression applies to the output $y(t)$. Equation (3.4) is the Volterra expansion of the dependence of state variable $u_n(t)$ on the input $x(t)$. The Volterra kernels $v_{nm}(s_1, ..., s_m)$ represent the mth degree dynamical dependence of $u_n(t)$ on $x(t)$.

If the system is reactive, then Eq. (3.4) applies and the relation of internal variable u_n with stimulus x as represented by v_{nm} can be investigated by means of correlation functions C_{nm} defined by

$$C_{nm}(s_1, ..., s_m) = \int dt\, u_n(t) \prod_{l=1}^{m} x(t - s_l). \tag{3.5}$$

The multivariate function C_{nm} represents the correlation of the state variable $u_n(t)$ with the mth-order product function of the stimulus $x(t)$. An essential requirement, usually not fulfilled by extracellular recordings, is that the state variable can actually be observed. In fact, the application of Eq. (3.5) requires simultaneous, intracellular multi-unit recordings. In the sequel we will proceed as if these recordings were already available.

In case the stimulus is taken as the realization $\underline{x}(t)$ of a Gaussian white process then the expected value of the functional C_{nm} is related simply to the Wiener kernels w_{nm}. For systems of known finite order $M(n)$ the set of Wiener kernels $w_{nm}(m = 1, M(n))$ is related unequivocally to the set of Volterra kernels $v_{nm}(m = 1, M(n))$ (Aertsen and Johannesma 1981a). However, accurate measurement and computation may be difficult if $M(n) \geq 4$. An important property of reactive systems is that each component $u_n(t)$ of the state variable $\boldsymbol{u}(t)$ can be analyzed separately with respect to its relation to the stimulus. Moreover Eq. (3.3) or more explicitly Eq. (3.4) indicate that the correlation between different components of the state variable is completely determined by the correlation of both components with the stimulus.

As a general conclusion it follows that reactive systems are solvable with respect to their characteristics and predictable with respect to their behaviour. The trajectories of their state variable are always regular and relax toward a single stable point in state space. In general, but not necessarily for each intermediate step, more reliable measurement and more elaborate computation will result in more precise knowledge concerning the system.

For creative systems the situation is quite different. For these systems we assert the following without proof (for reference, however, see Guckenheimer and Holmes 1983). The trajectory of the state variable $u(t)$ may, and probably will, show autonomous dynamics: attracted by one or more quasi-stable points or regions it may behave quasi-periodically or show chaotic behaviour determined by strange attractors. If parameters of the system change, even slightly, during the observation, catastrophic changes in behaviour may occur. Parameter estimation and trajectory prediction cannot be derived anymore from separate analysis of individual components $u_n(t)$ of state variable $\vec{u}(t)$; correlation between differents components u_n and u_m of the state variable is not primarily determined by their correlation with the stimulus. As a consequence the system should be analysed using specific model assumptions, based upon knowledge concerning development, structure and function of the system.

Some general remarks on the description of systems are relevant for the study of information processing in the nervous system. For any system three different aspects or levels are characteristic: *structure*, *process* and *function*. The structure of a system is the time invariant material base for the physico-chemical process associated with its internal dynamics and induced by influences from the environment (input) exerting again effect on this environment (output).

In this description reactive systems may also be named *heteronomous* systems: their behaviour is essentially determined by their input. On the other hand creative systems incorporate considerable *autonomous* aspects: their behaviour is only partly determined by the influences of the environment and the effect of this input depends on the present state of the system. In mathematical terms this follows from the irreducible form of Eq. (3.2). In physiological terms it implies the disappearance of the (stimulus invariant) receptive field.

In order to find a generalization for the receptive field for creative neural populations a fourth aspect only present in certain type of systems has to be considered. *Productive* or *throughput* systems which are evolved or designed primarily for the transformation of input into output, e.g. digestive system, are internally interesting mainly as far as there is relation with the input-output function. However, *cognitive* or *representational* systems, e.g. the visual system, not only have an input-output function, e.g. oculo-motor behaviour, but also create an internal image of the external world. For study of a representational system we should add *interpretation* as a fourth aspect to the system. This aspect is less objective than the other ones. Structure and process can be studied on a system in isolation, function is related to the interaction of system and environment, interpretation depends on the frame of reference of the observer. No interpretation can be given to a system of natural or artificial intelligence where there does not exist a sufficient overlap of (mental) images between subject and object. This does apply to computers, brains and languages.

The different aspects of a cognitive system are shown in Table 2.

Note that the present exposition was fundamentally based on the availability of adequate observations of the state variables u_n. As already stated this condition generally is not fulfilled in the context of electrophysiology. As such this framework should be regarded as an idealized view, important connections with experimental practice still are missing.

Table 2. Aspects of cognitive system

Interpretation	= Internal image of environment	Representation Transformation
Function	= Interaction with environment	Output Input
Process	= Physico-chemical activity	Variables Parameters
Structure	= Material substrate	Relations Elements

4 Neural Interaction

In order to relate these general notions on representational systems to the central nervous system and to the analysis and interpretation of multi-unit neural recordings a formal description is given for neural interaction. The neurodynamical equations for a neural population with time-invariant parameters are in a compact form given as follows (Johannesma and van den Boogaard 1985):

$$|U\rangle = |V\rangle + W * |z\rangle \tag{4.1a}$$

$$P(\langle z|\Delta t|U\rangle = \Lambda^{-1}\exp\langle z|U + R\rangle \tag{4.1b}$$

$$\Lambda = \sum_{\langle z|} \exp\langle z|U + R\rangle \tag{4.1c}$$

where

$\langle\cdot\|$	indicates a bra-vector
$\|\cdot\rangle$	indicates a ket-vector
$*$	indicates temporal convolution
$V_k(t) :=$	sensory input to neuron k at time t
$U_k(t) :=$	generating variable of neuron k at time t
$z_k(t) :=$	action-variable of neuron k at time t
$R_k :=$	spontaneous activity level of neuron k
$W_{kl} :=$	synaptic connectivity from neuron l to neuron k
$P_k(z_k\|\Delta t\|U_k) :=$	probability that neuron k generates an action-potential between t and $t + \Delta t$ given that its generator potential assumed the value U_k at time t.
$P(\langle z\|\Delta t\|U\rangle) :=$	probability that activity pattern at $(t, t + \Delta t)$ equals $\langle z\|$ given the state of the generator potential equals $\|U\rangle$ at time t.
$W :=$	connectivity matrix.

A sketch of the signal processing in two neurons according to Eq. (4.1) is given in Fig. 7. Linear spectro-temporal integration of action potentials in-

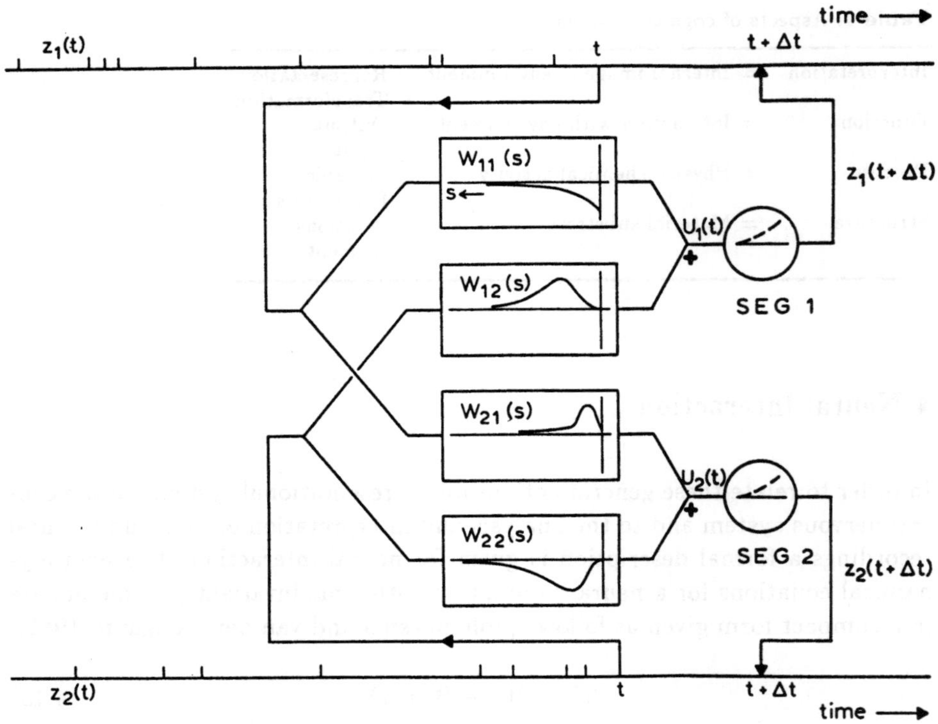

Fig. 7. Signal processing in neural interaction. Incoming sequences $z_l(t)$ of action potentials are temporally integrated and spatially summated into the generator potential $U_k(t)$ which again in a stochastic way creates the action potential $z_k(t+\Delta t)$. *SEG* indicates the stochastic event generator. (From Johannesma and van den Boogaard 1985)

duces the generator-potential, which in a local, nonlinear stochastic way again generates the action potentials.

Since the generator potential in neuron k exerts only an influence on the probability of spike generation in the same neuron it follows that the conditional probability of generation of an activity pattern factorizes in the product of conditional probabilities of the individual action potentials:

$$P(\langle z|\Delta t|U\rangle) = \prod_k P_k(z_k|\Delta t|U_k). \tag{4.2}$$

Under certain conditions for synaptic connectivity W the generator variable $|U\rangle$ may be considered as the state variable of the neural system forming the base of its dynamics. The partition function $\Lambda(|U\rangle)$, given by Eq. (4.1c), is mathematically analogous with the characteristic functional of a stochastic process and does allow the computation of all types of statistical properties of the neural population, including entropy, by proper differentiation. For more detailed mathematical treatment of these equations see Johannesma and van den Boogaard (1985), as well as van den Boogaard and Johannesma (1985).

In an equally compact form equations can be formulated for the development and/or plasticity of a neural population. In this point of view the connectivity matrix W is not longer a time invariant matrix of functions but as such the subject of a dynamical process, albeit on a longer time scale. The idea behind the equations is given by Hebb-like evolution of structure: temporal association induces structural connectivity. A general form for these equations can be given by

$$\frac{d}{dt}W(t,\tau) = \phi(\tau)\cdot\{|z(t-\tau)\rangle - Z(t-\tau|e\rangle\}\{\langle z(t)| - \langle e|Z(t)\} \qquad (4.3)$$

where:

$W :=$ matrix of connectivity functions
$\phi :=$ scalar evolution kernel
$\langle z| :=$ neural activity pattern
$\langle e| :=$ unit vector $= \frac{1}{\sqrt{K}}(1,1,1,...,1)$

$$Z := \langle e|z\rangle = \frac{1}{\sqrt{K}}\sum_{k=1}^{K} z_k$$

Z forms a measure for total or average actitivity of the population.

In Eq. (4.9) two constraints have been incorporated:

$$\frac{d}{dt}\sum_k W_{kl} = 0 \quad \text{or} \quad < e|W \quad \text{is time-invariant}$$

which implies that the sum of output weights is invariant preventing absolute domination of a single neuron on the population.

$$\frac{d}{dt}\sum_l W_{kl} = 0 \quad \text{or} \quad W|e\rangle \quad \text{is time-invariant}$$

which implies that the sum of input weights is invariant, preventing absolute submission of a given neuron to the population.

If $\phi(\tau)$ is positive the plasticity develops in the form of cooperative evolution, for negative $\phi(\tau)$ a competitive evolution takes place. Combined forms can be represented by e.g. a $\phi(\tau)$ which is positive for small τ and negative for large τ. Equation (4.1) for neural interaction and Eq. (4.3) for synaptic evolution have a long history in brain theory (Caianiello 1961). The form of the connectivity matrix W determines the behaviour of the neural population in a fundamental way. In our opinion the following property holds: if W is not in triangular form and cannot be transformed in triangular form by a simple permutation of the neurons, then the neural population has creative features and cannot be analysed purely in terms of receptive and effective fields of its neurons. This implies that an evolution of W, according to Eq. (4.9), from a triangular form to a non-triangular form may transform the neural population from reactive into creative and change its information processing in an essential way.

5 Conceptual Base for Evaluation of Neural Activity Patterns

For a systematic approach to analysis and interpretation of neural activity patterns a fundamental hypothesis is formulated:

Neural activity patterns

- are evoked by sensory stimuli (receptive field)
- transformed by neural populations (assemblies)
- specified by sensitivity and connectivity (structure)
- as evolved in phylogenesis (selection)
- and modified in ontogenesis (plasticity)
- in such a way as to enhance forms of neural synchrony (coherence)
- which induce perceptual integration (harmony)
- and produce behavioural coordination (synergy)

A diagram of the mechanisms and influences expressed in this hypothesis is given in Fig. 8.

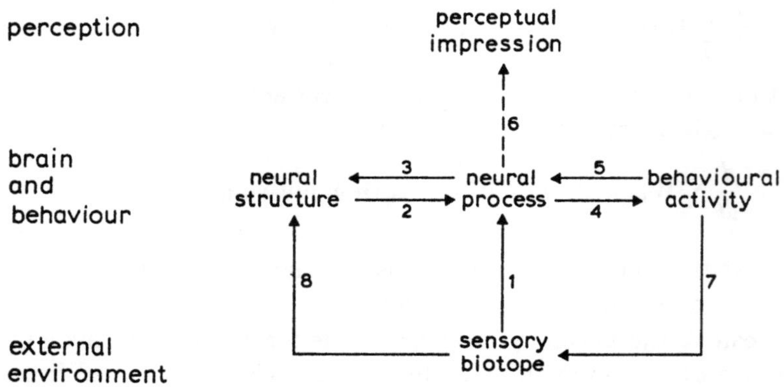

Fig. 8. Mechanisms and interactions influencing the activity patterns in a neural population

The nature of the interconnections, indicated by arrows in Fig. 8 is as follows:

1. Influence of sensory stimulus on neural activity ($10^{-3} - 1$ s.)
2. Influence of neural structure (synaptic connectivity) on neural activity patterns ($10^{-3} - 1$ s.)
3. Influence of neural activity on structure of the population: plasticity, associative memory, "learning" ($1 - 10^3$ s.)
4. Neural production of behavioural activity ($10^{-3} - 1$ s.)
5. Neural sensitivity for behavioural activity: proprioception ($10^{-3} - 1$ s.)

6. Sensory interpretation of neural activity ($10^{-3} - 10^{3}$ s.)
7. Influence of behaviour on sensory stimuli and sensory biotope ($10^{-3} - 10^{3}$ s.)
8. Evolutionary selection of neural structures ($10^{5} - 10^{8}$ s.)

A consequence of this hypothesis is the definition of a neural assembly as a population of neurons which generates a coherent neural activity pattern. Coherence of neural activity should then be recognized and understood on the base of its functional interpretation as specified by the relation to internal structure of the sensory biotope (symphony), to perceptual integration (harmony) and behavioural coordination (synergy). An illustrative drawing of this point of view is given in Fig. 9.

In this line of thought a neural assembly is a dynamic structure influenced both by external variables and internal states; however, the time constant of variation of composition of assemblies is assumed to be considerably longer than the time constant involved in elementary aspects of perception and behaviour. The "harmony" in perception can be related to the tendency to favour sensory impressions with natural plausibility and relevance. The "synergy" in behaviour is related to elegance and efficiency of behavioural sequences. "Symphony", "synergy", "harmony" and "assembly" are considered as functions on their respective domain of definition. The mappings of these functions are assumed to be simpler than the mappings of their arguments.

The inclusion of the aspect "perception" in Fig. 8 and "mind" in Fig. 9 is an attempt to include the "representational" aspect of the nervous system. It is not intended as a description of experimental data, but as a conceptual interpretation. Mental aspects of neural activity are not operational variables, but formal expressions of these variables derived from meta-physiological consid-

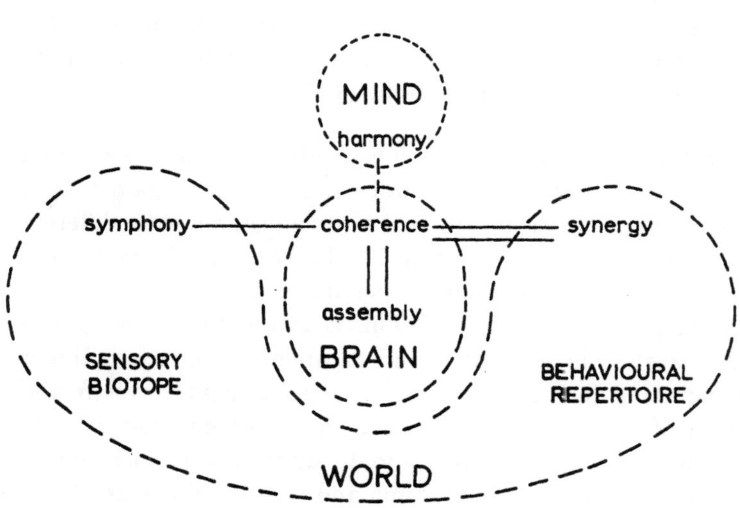

Fig. 9. Sketch of sensory "symphony", behavioural "synergy" and mental "harmony" in their relation to neural "assembly"

erations e.g. evolution and behaviour. From the point of view of experimental neurophysiology this may appear speculative, on the other hand it may well be that the specific value of stimulus associated multi-unit recording compared to single unit recording is given by the data it supplies with respect to the neural base of perception and behaviour. The multi-unit recording gives a specific possibility to investigate the creative processes of neural populations and as such it appears to be connected to the representational function.

An illustration of these ideas results from the study of different types of animal behaviour; e.g. predation, territorial and sexual behaviour. In these situations a specific sequence of behavioural acts occurs which has been selected out of a set of possible behaviours directed towards certain goals and based upon interpretation of the sensory environment and expectation of its development. If a frog attempts to catch a flying insect he has to match his intended jump to the extrapolation of the trajectory of the insect. The explanation of the actual behaviour of the frog can only be given by an embedding of these facts among the possible alternative forms of behaviour e.g. a different type of jump at a different moment of time. This selection of behaviour results from the characteristics of an active model of the environment carried by a creative neural system forming a representation of the perceptual-behavioural interaction of the animal and its environment.

It is in this direction that we attempt to find an approach to analysis and interpretation of multi-unit recording considered as an observation of neural activity patterns.

6 Analysis and Interpretation of Neural Activity Patterns

The fundamental assumption is that the analysis and interpretation of stimulus associated multi-unit recording should be based on the description of the brain as a creative representational system generating an internal model of the perceptual and behavioural environment present in the past and persistent in the future.

The experimental data given by multi-unit recordings form an observation of the nervous system on the level of the process. In order to gain insight in the nervous system as such we try to make a connection with different levels of the system. One of these connections is the relation of process and structure: how to go from the correlation measured in the neural activity to the connectivity present in the neural structure? This is an inverse problem, well-known in the field of system identification and parameter estimation. Under certain conditions, including qualitative information concerning the system and complete observability of the state variables, it may be possible to compute the behaviour of the system in a forward way and to invert these equations in order to determine the parameters (Johannesma and van den Boogaard 1985, van den Boogaard et al. 1985). A more inspired approach has been taken by Gerstein; in a paraphrase it can be stated as follows:

1. Make an assumption on the ontogenesis of the system in relation to its process.
2. Model the system in a qualitative form.
3. Observe the activity patterns belonging to this system.
4. Project a development on the model based on the ontogenetic principles in combination with the observed processes.
5. Assume that the asymptotic structure of the time-variant model reflects the time-invariant structure of the neural population.

Compare the contribution of Aertsen et al. in this Vol.

This approach as a form of adaptive system identification is interesting and looks promising. Based on the ideas of different levels of systems, including the nervous system, two observations can be made.

In the first place all anatomical and physiological information available a priori should be included in the initial abstract form of the model. If equations of the type of Eq. (4.3) do apply, they may well be used, moreover then the initial condition on W should not be $W(0) = 0$ or random but based on assumptions and observations e.g. using the (relative) positions of the microelectrodes.

In the second place the view of the nervous system as a representational system implies that the model fo the nervous system as generated in this procedure forms a model of reality. If the model of reality, the brain, has developed in a natural environment of sensory stimuli and behavioural patterns then it may be preferable to have the development of the model of the brain occur in a comparable situation. This leads to the conclusion that natural stimuli and active behaviour may well be a necessary condition for a successful imitation of the ontogenesis of a neural population.

In these considerations we see an interaction of the syntactical analysis of a neural population (structure ↔ process) and the semantical level (process ↔ function). This idea also appears fundamental to the notion of a *neural assembly*: a population of neurons connected in such a way (structure) that their activity pattern (process) corresponds to a meaningful element in the sensory-motor space (function) (Braitenberg 1977, Palm 1982). If this description, as given in Sect. 5, is accepted, then the study of a neural assembly should be made as a multi-level investigation of structure, process and function of neural populations.

The "gravitational clustering" approach starts from the syntactic level concentrating on neural interaction and trying to deduce connectivity (structure) from correlation (process); in a later stage stimulus related arguments (function) become relevant. Now we shall sketch a complementary approach starting on the semantic level as given by the relation of sensory stimulus and neural activity. In fact the ideas are focused upon the neural base of perception.

Apart from introspection, the perceptual value of sensory stimuli can be measured by behavioural responses in relation to sensory stimuli. The normal procedure is the presentation to a subject, animal or human, of pairs of sets of stimuli, which are selected in such a way that these can barely be discriminated among each other and ask or force the subject to make the discrimination. This leads to the "just noticeable difference" (j.n.d.) for sensory stimuli; these j.n.d.'s can be used as a metric in sensory space. The sensory space together with the metric or resolution from the j.n.d. has the mathematical form of a "tolerance-space" and may be considered as the perceptual space. The characteristics of the perceptual space may well depend on the situational context of the experiment and on the motivational state of the subject. However, giving confidence to experimental psychology and/or ethology we assume this perceptual space to be known in principle.

A central question for the interpretation of neural activity in sensory parts of the nervous system is the definition, both formal and operational, of the perceptual value of neural activity. We propose the following definition:

> the percept associated with a neural activity pattern is the optimal estimation of the sensory stimulus which can be made by an external observer which has knowledge both of the structure of the nervous system and of the characteristics of the stimulus ensemble.

Here the notion of "optimal" is defined as a least square distance of percept and stimulus in technical context and as a minimal perceptual distance in a behavioural context. The perceptual metric has to be based on meta-physiological information e.g. from behavioural studies. This definition is analogous to the situation of an observer who knows his apparatus and the general characteristics of his environment and then searches for the best interpretation of his observations.

The percept as defined here is not a physical entity as such but an interpretation by the investigator projected upon his experimental data in the context of his insight into the behavioural relevance. When the neural activity pattern forms the internal map of the sensory stimuli, then the percept is defined as the optimal inverse map from neural activity back to the sensory stimulus. If the mathematical description of the nervous system, as far as involved in this sensory task, can be formulated as well as the description of the sensory environment, then the inverse map can in principle be derived making use of the Bayes' relation for conditional probabilities. For a reactive model of the peripheral part of the auditory system this has been investigated mathematically and by computer simulation (Johannesma 1981).

The percept associated with a given neural activity pattern z is denoted as the sensory interpretation of neural activity (SINA) and in first approximation defined as the appropriate superposition of single unit receptive fields (SURF) of those neurons which are active. In general receptive fields, visual and auditory, will be spatio-spectro-temporal entities which may be regarded as points or regions in sensory space. For a reactive neural population the SURF is a properly defined entity which can be measured e.g. by cross-correlation of sensory stimulus and single unit neural events. It may be clear that the SURF is described in sensory space (adequate or optimal stimulus or stimuli to activate a neuron); as such it forms already an example of inverse mapping. The SINA is intended as a generalisation of the SURF for multi-unit activity. However, as far as our insight goes, the SURF is only defined for a reactive population, while the SINA, being intended for the functional description of a neural assembly, is related to a creative population. A solution for this vicious circle is a sequential approximation under the assumption of convergence to an appropriate asymptotic solution. The verification of this assumption needs further mathematical analysis, numerical simulation and application to electrophysiological data.

Before the precise description of the SINA a final assumption is made concerning the interrelation of structure, process and function in the nervous system. It is based on the ideas presented in Sect. 4 and the definition of reactive and creative as given in Sect. 3. The assumption is as follows:

- Reactive aspects of the nervous system (convergence and divergence of connections) take care of the ***representation*** of the sensory environment directed toward ***precision*** and ***reliability*** (signal/noise).
- Creative aspects of the nervous system (lateral and recurrent connections) take care of the ***transformation*** of the representation of the sensory environment directed toward ***plausibility*** and ***relevance*** (figure/ground).
- The interplay of reactive and creative operation is under control of central parts of the nervous system which are influenced by external and internal factors.
- Reactive aspects are relatively insensitive, creative aspects are relatively sensitive for composition of the stimulus ensemble and state of the animal: the more natural the state of the animal and the characteristics of its environment are, the more creative the nervous system behaves.

On this base a description in formal and a prescription in operational terms of the SINA reads as follows:

1. Induce a perceptually alert state in the animal under investigation.
2. Present in alternating way two stimulus ensembles:
   the first one with form but without sense:
   e.g. random elements of the sensory biotope in random order
   the second one with form and with sense:
   e.g. natural selections of the sensory biotope.
3. Measure the stimulus associated multi-unit neural activity pattern $\{z_k(t), k = 1, K\}$ for both stimulus ensembles.
4. Compute the SURF for the first stimulus ensemble by correlation of sensory stimulus and neural events:
   e.g. by second order correlation for the auditory system
   $$X_k(\mathbf{r},\omega,\tau) = \int dt\, X(\mathbf{r},\omega,t-\tau) z_k(t)$$
   where $X(\mathbf{r},\omega,t)$ is the spatio-spectro-temporal characterisation of the stimulus ensemble.

5. Construct the SINA for both stimulus ensembles by an appropriate superposition of the receptive fields of the active neurons

$$\Xi(\boldsymbol{r},\omega,t) = \sum_{k=1}^{K} \int d\tau \, X_k(\boldsymbol{r},\omega,\tau) z_k(t-\tau)$$

6. Display the spatio-spectro-temporal image of stimulus-ensembles presented to the animal as well as the image of both stimulus-ensembles as constructed on the base of the associated neural activity pattern.
7. Investigate the hypothesis that the SINA of the sensible S.E. is transformed compared to the SINA of the non-sense S.E. in the direction of greater realism; the figure-ground effect should be much stronger.

A comparable experiment would be to choose a single stimulus ensemble where the stimuli gradually change from a given natural stimulus through an intermediate stage to a different natural stimulus. In this situation we expect a phase-transition in the neural activity pattern such that the associated SINA abruptly changes from one percept into the other; also hysteresis type of effects may occur.

A neural assembly may then be defined as a creative population of neurons which contribute to the generation of

- figure out of ground
- percept out of stimulus
- sense out of fact.

Acknowledgements. This investigation forms part of the research-program "Brain and Behaviour" of the Department of Medical Physics and Biophysics, University of Nijmegen and was supported by the Netherlands Organization for Advancement of Pure Research (ZWO). The manuscript was prepared by Marianne Nieuwenhuizen.

References

Aertsen AMHJ, Johannesma PIM (1981a) The spectro-temporal receptive field. A functional characterization of auditory neurons. Biol Cybern 42:133–143

Aertsen AMHJ, Johannesma PIM (1981b) A comparison of the spectro-temporal sensitivity of auditory neurons to tonal and natural stimuli. Biol Cybern 42:145–156

Boogaard HFP van den, Johannesma PIM (1984) The Master Equation for Neural Interaction. IMA J Math Appl Med Biol 1:365–389

Boogaard HFP van den, Hesselmans GHFM, Johannesma PIM (1985) Transformation of point processes, correlation functions and system identification. Submitted

Braitenberg V (1977) Cell assemblies in the cerebral cortex. In: Hein R, Palm G (eds) Theoretical approaches to complex systems. Lecture notes in biomathematics, vol 21. Springer, Berlin Heidelberg New York, pp 171–188

Caianiello ER (1961) Outline of a theory of thought processes and thinking machines. J Theor Biol 1:209

Eggermont JJ, Aertsen AMHJ, Johannesma PIM (1983a) Quantitative characterisation procedure for auditory neurons based on the spectro-temporal receptive field. Hearing Res 10:167–190

Eggermont JJ, Epping WJM, Aertsen AMHJ (1983b) Stimulus dependent neural correlations in the auditory midbrain of the grassfrog (Rana temporaria L.). Biol Cybern 47:103–117

Eggermont JJ, Johannesma PIM, Aertsen AMHJ (1983c) Reverse correlation methods in auditory research. Q Rev Biophys 16:341–414

Epping W, Boogaard H van den, Aertsen A, Eggermont J, Johannesma P (1984) The neurochrome: an identity preserving representation of activity patterns from neural population. Biol Cybern 50:235–240

Gerstein GL (1970) Functional association of neurons: detection and interpretation. In: Schmitt FO, Quarton GC, Melnechuk T, Adelman G (eds) The neurosciences: 2nd Study Program. Rockefeller Univ Press, New York, pp 648–661

Gerstein GL, Michalski A (1981) Firing synchrony in a neural group: Putative sensory code. In: Székely G, Lâbos E, Damjanovich S (eds) Neural communication and control. Adv Physiol Sci vol 30. Pergamon Press, Akad Kiado, Budapest, pp 93–102

Guckenheimer J, Holmes P (1983) Nonlinear oscillations, dynamical systems and bifurcations of vector fields. Springer, Berlin Heidelberg New York

Hebb DO (1949) The organization of behavior. A neuropsychological theory. Wiley, New York; Chapman and Hall, London

Johannesma PIM (1972) The pre-response stimulus ensemble of neurons in the cochlear nucleus. In: Cardozo BL (ed) Proc IPO Symp Hearing Theory, Eindhoven

Johannesma PIM (1980) Functional identification of auditory neurons based on stimulus-event correlations. In: Brink G van den, Bilsen FA (eds) Psychophysical, physiological and behavioural studies in hearing. Delft Univ Press, pp 77–84

Johannesma PIM (1981) Neural representation of sensory stimuli and sensory interpretation of neural activity. Neural communication and control. Adv Physiol Sci 30:103–125

Johannesma PIM, Aertsen AMHJ (1982) Statistical and dimensional analysis of the neural representation of the acoustic biotope of the frog. J Med Syst 6(4):399–421

Johannesma PIM, Boogaard HFP van den (1985) Stochastic formulation of neural interaction. Acta Math Appl 4:201–224

Johannesma P, Eggermont J (1983) Receptive fields of auditory neurons in the midbrain of the frog as functional elements of acoustic communication. In: Ewert JP, Capranica RR, Ingle DJ (eds) Advances in vertebrate neuroethology. Plenum Publ, New York, pp 901–910

Palm G (1982) Neural assemblies. An alternative approach in artificial intelligence. Studies of brain function, vol VII. Springer, Berlin Heidelberg New York

Scheich H (1977) Central processing of complex sounds and feature analysis. Recognition of complex acoustic signals. In: Bullock TH (ed) Life Sci Res Rep, vol V. Springer, Berlin Heidelberg New York

Snyder DL (1975) Random point processes. Wiley, New York

On Information Processing in the Cat's Visual Cortex

W. VON SEELEN, H.A. MALLOT, G. KRONE, and H. DINSE[1]

1 Introduction

We assume that the visual system serves for orientation in space, recognition of objects and the interpretation of scenes and scene sequences. This task breaks up into a series of partially interdependent subproblems which are solved by some 13–15 usually retinotopically organized areas. So far it has not been possible to correlate functions and areas unequivocally. One reason for this could be the inadequacy of the questions posed as a basis for experiments. However, we think it more likely that correlating a function with an area is, as a rule, inadmissible since the degree of the coupling in the whole system does not permit a simple divsion. Rather the type and degree of coupling determine the separation or integration of "elementary units". It follows that co-operation between subsystems is an essential aspect. The data available to us from neurophysiological, neuroanatomic and behavioral physiological findings is insufficient to analyse such a system: they ensure neither the complete observability nor the complete controllability of the systems. So one is forced to use model comparison for analysis. Depending on the type of data, we use three models which complement each other to order the data and predict results which can be experimentally verified.

The first model (Sect. 2) simulates the recognition of simple patterns and is used for studying which features are extracted, which operation divides feature space and how areas 17, 18, and 19 contribute to this task. The experimental method includes measuring the animal's performance in distinguishing a pattern from additive visual noise.

The second model (Sect. 3) employs a synthetically oriented approach: a cortex model with six layers based on the neuroanatomic structure of the cortical areas was implemented involving a number of coupling principles we consider important. Its purpose is to analyze the fundamental, layer specific characteristics of such systems, in particular their dynamics. Both the neuroanatomic and the neurophysiological data came mainly from area 17. This

[1] Institut für Zoologie III (Biophysik), Johannes Gutenberg-Universität, Saarstr. 21, 6500 Mainz, FRG

Brain Theory
Edited by G. Palm and A. Aertsen

study was carried out in co-operation with G. Palm and A. Schüz of the Max-Planck-Institute for Biological Cybernetics, Tübingen.

The third model (Sect. 4), closely connected to the second, takes more particular account of an overall aspect: the analysis of the functional consequences of neuronal mapping in areas 17, 18, and 19. This organizational aspect seems essential to us since alterations to the internal coordinate systems result in considerable computational advantages for various tasks.

The three models emphasize different aspects of the same problem: how do the subsystems of the visual cortex cooperate in performing different tasks?

Several fundamental assumptions are involved in the details of the experiments and models. First it is assumed that the "basic system" for the models is the two-dimensional, laterally coupled neuronal layer. It is also assumed that any lesion in the system reduces its performance, i.e., that the precision and certainty of the systems are interdependent. It follows that "redundancy" is made use of, even in cases where there is no disturbance, i.e., there is no real redundancy at all. The third assumption is that the system is time variant, i.e., the functional structure can be varied for a particular task depending on the external stimulus situation. Simple experimental situations can easily lead to simple models well capable of providing predictions although they represent only a partial aspect of more complex tasks. These assumptions, which cannot be proven at present, will be used in what follows in various ways.

2 On the Analysis of the Cat's Pattern Recognition System

Our work attempts to answer two questions:

1. Which operations suffice to describe pattern recognition?
2. What do areas 17, 18, and 19 contribute to this task?

The experiments consisted in binary detection tests in which the cat has to distinguish between two patterns with superimposed visual noise. The probability of correctly detecting the pattern, $P_D = f(S/N)$, constitutes the measurable variable and the efficiency measure. If the pattern is interpreted as a point in n-dimensional space, then the visual system transforms this pattern into an internal image which contains the essential features, but has less dimensions than the original as a rule. This stage of feature extraction is followed by classification, i.e., the division of feature space into classes of significance. On the basis of this interpretation, behavioral experiments are mostly carried out in such a way that the experimenter varies the stimuli until one of the internal representations exceeds the separation plane and a false classification occurs.

The drawback in this procedure is the large number of possibilities open to complex visual systems and consequently the high experimental effort involved. For this reason we adapted the basic principle of systematically exceeding the separation plane by using visual noise and the internal noise of the system so that the probability of false classification can be calculated (it is mainly from false classifications that conclusions on the system can be drawn). The experimental findings were used to construct a model and estimate its parameters. This model is based on the total of performances of intact animals.

It is impossible to determine the individual contributions of area 17, 18, or 19 to pattern recognition when the system is coupled, and it cannot be decoupled experimentally in any simple way. So lesions were made on the areas, and the animal's performance in the relevant task was ascertained before and after the operation. Since it can be assumed that, due to their coupling, the areas can partially replace each other, the performance deficiency measured has to be interpreted as the inability of the remaining system to compensate for the defect caused by the lesion. The number of lesions required depends on the coupling of the areas and on the number of coupling functions to be determined.

2.1 Experiments

The results relevant to the model obtained from four types of lesion are shown in Table 1. The task tested for this model was that of distinguishing between two patterns (circle, triangle) on which a visual noise process was superimposed. This spatial noise – an on-axis Fourier hologram of a piece of ground glass – had a normal amplitude distribution and a constant spectral power density. The variable to be measured was always the detection probability P_D as a function of the signal to noise ratio. The parameters varied in the experiments were the size of the patterns, their movement relative to the noise, and the bandwidth of the noise. The four lesion groups in Table 1 each consisted of three animals which were first trained and tested, then lesioned, and then tested again. The quality of the lesions was checked histologically. The experiments are discussed in more detail by Krüger et al. (1986).

Table 1. Task: DISCRIMINATION of 2 PATTERNS

Sig.	Noi. \ Les.	17	17+18	19	17+18+19
M	N		×	×	×
M_V	N		×	×	×
m	N	×	×	×	×
m_V	N	×	×	×	×
m	N_V			×	
m_V	N_V	×		×	×
m	N_F		×		
m_{V_i}	N	(×)		×	

M = large Patterns
m = small Patterns
M_V, m_V = moved Patterns
N_V = moved Noise
N_F = filtered Noise
m_{V_i} = moved Patterns with Velocities Vi

$P_D = f(s) / \frac{S}{N} = \text{Const.}$

2.1.1 Experiment A

Three intact animals, experimental parameter: movement of the signal (V_m) and of the noise (V_n). The results are shown in Fig. 1. As one can see, the increase in P_D is small when the patterns move, which is quite in contrast to the results of similar experiments on humans.

2.1.2 Experiment B

Three animals, lesion of areas 17 and 18, experimental parameter: size of patterns in the stationary case. Figure 2 shows the performance of the animals in the test situation. After the lesion, there is only a slight deficit for large patterns, while the decrease in performance is considerable for small patterns.

2.1.3 Experiment C

Three animals, lesion of area 17, experimental parameter: size of the patterns in the stationary case.

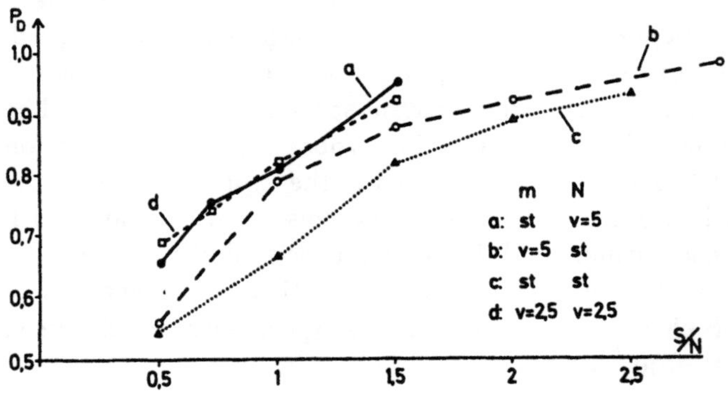

Fig. 1. Detection curves for varios velocities. *m* signal; *N* noise; *v* velocity (cm s^{-1}); *st* stationary

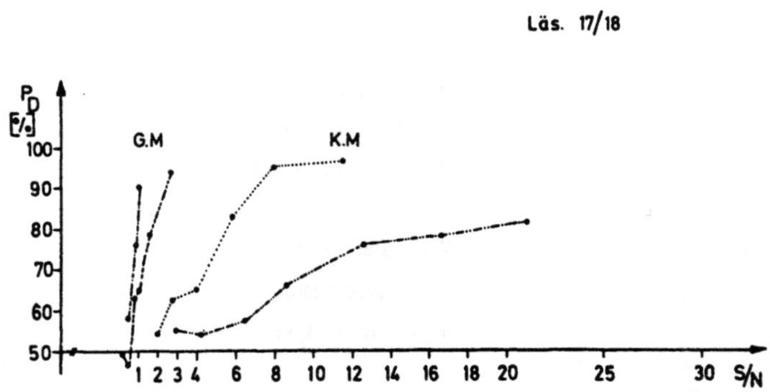

Fig. 2. Performance of the cat (lesion of areas 17 and 18). *LN* large patterns, normal cat; *LD* large patterns, lesioned cat; *SN* small patterns, normal cat; *SD* small patterns, lesioned cat

Figure 3 shows that the detection performance corresponds to that achieved after lesion of area 17/18. By contrast, comparing the effects of the two lesions on the detection of moving patterns shows that a lesion of area 17/18 elicits a distinctly larger deficit. However, the results relating to area 18 require further experimental confirmation since the area 17 lesion caused some damage to area 18 as well.

2.1.4 Experiment D

Three animals, lesion to area 19, experimental parameters:

a) $V_m = 0, \ V_n = 0$
b) $V_m = V_i \quad i = 1, \ldots, 4, \ V_n = 0$
c) $V_m = V_n = 2.5\,\mathrm{cm\,s^{-1}}$

As can be seen in Fig. 4, which shows the difference in detection performance before and after the lesion for the case a), there is no decrease in performance. Differences are detectible only in the lower velocity range (Fig. 5).

2.1.5 Experiment E

Three animals, lesion to areas 17, 18 and 19; variable to be measured: P_D for the pattern without noise; experimental parameter: pattern size, $V_m = V_n = 2.5\,\mathrm{cm\,s^{-1}}$.

Figure 6 shows that the animals are incapable of carrying out the task set in the three given experimental situations. This experiment shows that the midbrain path circumventing the three cortical areas can make only slight contributions.

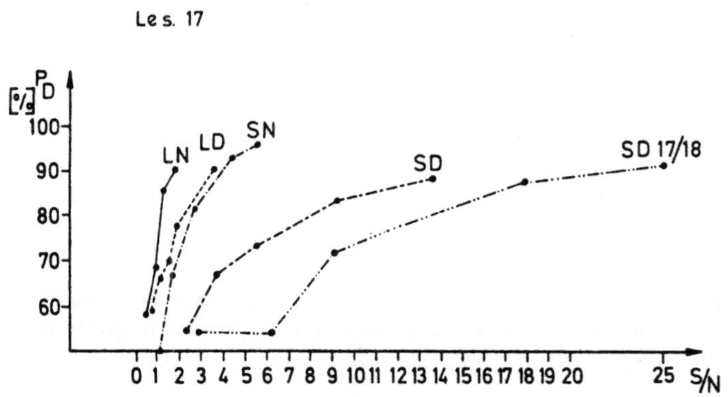

Fig. 3. Performance of the cat. *LN* stationary large patterns, normal cat; *LD* stationary large patterns, lesioned cat (area 17); *SN* stationary small patterns, normal cat; *SD* stationary small patterns, lesioned cat (area 17); *SD17/18* moving small patterns, lesioned cat (areas 17 and 18)

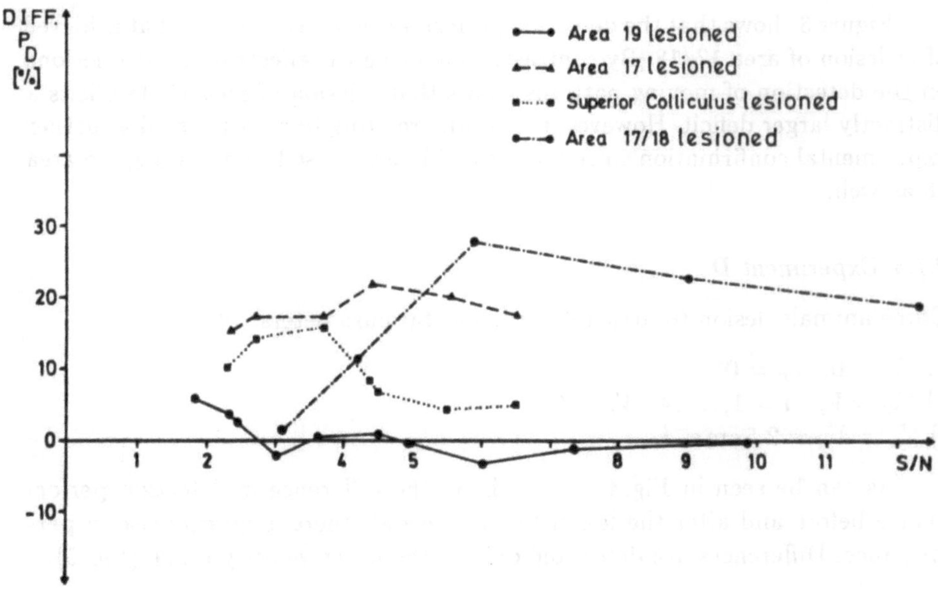

Fig. 4. Performance difference between normal and lesioned cats

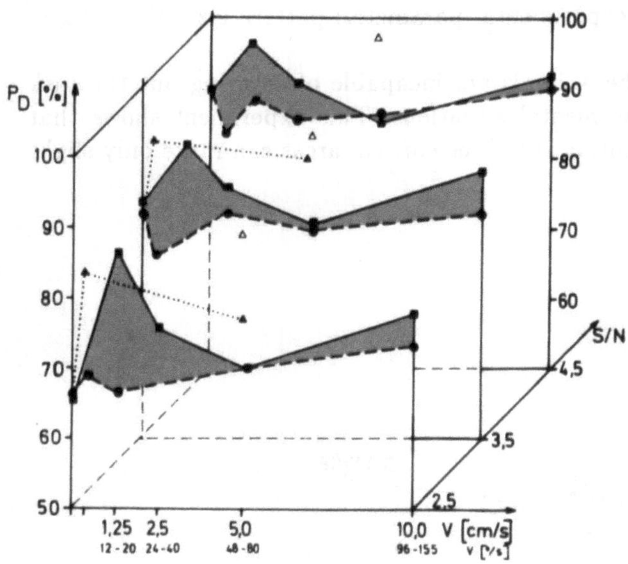

Fig. 5. Performance difference between normal and lesioned cats (area 19). Moving pattern, stationary noise: □ — □ before lesion; • - - - • after lesion. Moving noise, stationary pattern: △ ... △ before lesion; ▲ after lesion

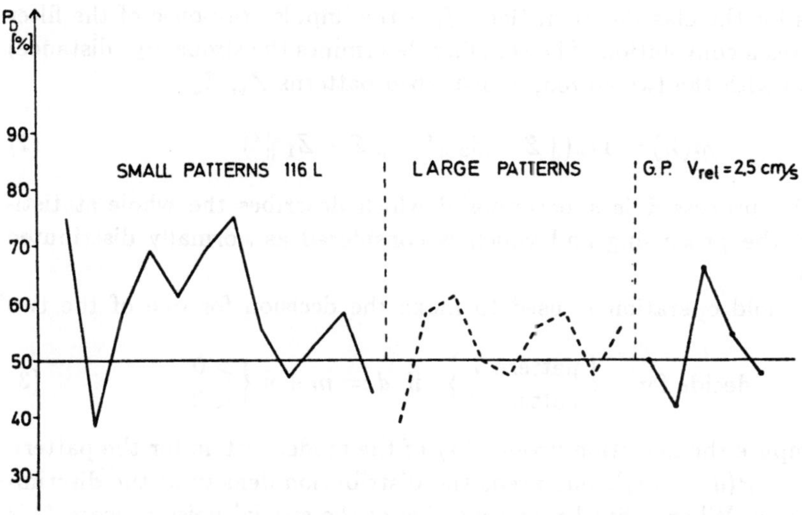

Fig. 6. Performance after lesion to areas 17, 18, and 19

2.2 Model

The experiments were quantitatively evaluated with a model (Fig. 7), the structure of which was suggested partly by the experiments described and partly by earlier studies. It assumes that the recognition system is divided into a feature extraction and a classification part. The first system involves the acquisition and transformation of the disturbed signal $x = \alpha S_1 + \beta N$ or $x = \alpha S_2 + \beta N$ by the retina (F_1), a subsequent normalization of the energy of the resulting signal to unity and a further filtering (F_2). Filter 1 is a linear system, the transfer function of which corresponds to the curve of contrast sensitivity as measured by Campbell et al. (1973) rotated around the zero point. Filter 2 is the system which is determined by matching the simulations to the experimental data; here the lesions are supposed to interfere. The signal obtained in this way,

$$Z = f_2 * f_1 * X / \parallel f_1 * X \parallel , \tag{1}$$

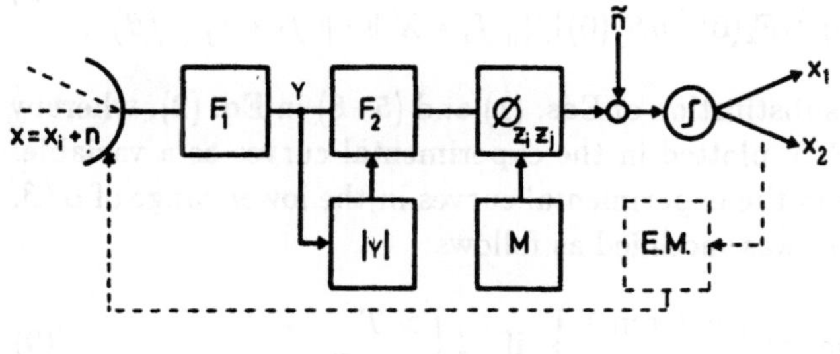

Fig. 7. Model of the recognition system (for details see text)

is the basis for the classification. Here, f_i is the impulse response of the filter i; "$*$" denotes a convolution. The classifier determines the similarity (distance) of the signal with the two stored, undisturbed patterns Z_1, Z_2 :

$$m(Z) = 1/2(\| Z - Z_2 \|^2 - \| Z - Z_1 \|^2) \ . \tag{2}$$

Then a noise process $\tilde{n}$ is superimposed which describes the whole statistical error in the processing and which is considered as normally distributed ($\sim N(0,1)$).

A threshold operation is used to make the decision for one of the two classes:

$$\text{decide for} \quad \begin{Bmatrix} \text{pattern 1} \\ \text{pattern 2} \end{Bmatrix} \quad \text{if } d := m + \tilde{n} \begin{cases} > 0 \\ < 0 \end{cases} \tag{3}$$

To compute the detection probability of the model system for the pattern S_1, $P_D(S_1) = P(d > 0|S_1)$, one needs the distribution density of the discrimination value d. When a fixed representative of the spatial noise process N is used, m is no longer a random variable. The distribution of d is then normal with the mean m and variance 1 so that

$$P_D = \Phi(m(Z_1)) \ , \tag{4}$$

where Φ denotes the error function.

The variable to be computed is the mean m from Eq. (2) (for a detailed derivation of these results cf. Krone et al. 1983). We substitute Eq. (1) in Eq. (2) and compute the required expressions:

$$\begin{aligned} \| f_1 * X \|^2 = & \alpha^2 \| f_1 * S_1 \|^2 + \beta^2 \| f_1 * N \|^2 \\ & + 2\alpha\beta |(f_1 * S_1, f_1 * N)| \end{aligned} \tag{5}$$

$$\| f_1 * N \|^2 = \int \varphi(\omega)|F_1(\omega)|^2 d\omega + \mu^2 |F_1(0)|^2 \tag{6}$$

$$(f_1 * S_j, f_i * N) = \mu |F_1(0)|^2 S_j(0) \tag{7}$$

with F_i as the Fourier transform of f_i and μ as the mean and $\varphi(\omega)$ as the power density spectrum of the spatial noise process.

$$\begin{aligned} (Z, Z_j) = & [\alpha/\beta \int |F_1(\omega)|^2 |F_2(\omega)|^2 S_1(\omega) S_j(\omega) d\omega \\ & + |F_1(0)|^2 |F_2(0)|^2 \mu S_j(0)] / (\| f_1 * X \| \cdot \| f_1 * S_j \| / \beta) \ . \end{aligned} \tag{8}$$

The result is obtained by substitution of Eqs. (1) and (5–8) in Eq. (2), whereby α/β corresponds to the S/N plotted in the experimental curves as a variable.

Due to the lower rise in the experimental curves in the lower range of α/β, the classification operation was modified as follows:

$$\text{decide for} \begin{Bmatrix} \text{pattern 1} \\ \text{pattern 2} \end{Bmatrix} \quad \text{if } d \begin{cases} > T \\ < -T \end{cases} \tag{9}$$

If $-T \leq d \leq T$, then a random decision with a probability of 0.5 is made for one of the two classes. It follows that:

$$P_D(S_1) = \Phi[m(Z_1) - T] + 0.5\{\Phi[m(Z_1) + T] - \Phi[m(Z_1) - T]\} \ . \tag{10}$$

The results of two kinds of stationary case experiments (detection of large patterns, and detection of small patterns) for each of the four lesion types were used to estimate the filter F_2, which was therefore chosen so that two parameters were to be determined:

$$F_2(\omega_1, \omega_2) = A \cdot (\omega_1^2 + \omega_2^2) exp[-B^2(\omega_1^2 + \omega_2^2)] \ . \tag{11}$$

The model simulations, which consist in an estimation of the parameters A and B of the Filter F_2, were carried out for the stationary case alone. The relevant experiments can be summarized in two cases:

Case 1: recognition of large and small patterns by cats without lesions
Case 2: recognition of large and small patterns by cats with lesion to areas 17 and 18.

Since, as Fig. 4 shows, there is no significant difference between pre- and post-operative performance for animals with lesion to area 19, Case 1 covers the lesion to area 19 with respect to the recognition of small patterns. For the same reason (cf. Sect. 2.1.4, Experiment D) lesion to area 17 is covered by Case 2, i.e., the lesions of area 17 and area 17/18 are not distinguished. The curves in Fig. 9 contain the results from all experiments for the stationary case. The simulation results for these situations with a fixed threshold $T = 0.5$ are:

Case 1: $F_2 : A = 0.001, \quad B = 3.46°$
Case 2: $\tilde{F}_2 : A = 0.0013, \quad B = 4.25°$

The resulting filter curves are shown in Fig. 8, where Filter 1 and the frequency characteristic of the small circle are also plotted for comparison. The model with the specified parameters leads to a performance efficiency such as is shown in Fig. 9 (the experimental values are plotted as points). The model was tested with another simulation with a pattern set not used as a basis for parameter estimation, viz., vertical and horizontal gratings. The solid line in Fig. 9 shows the reliability of the predictions of the model; the points marked with crosses are the measured values.

The simulation results make it clear that the lesions lead to a remainder system with a reduced bandwidth and lower amplitudes compared to those of the normal system. The narrow bandwidth of the filter is striking and results from the fact that the system parameters were estimated only in the frequency range in which the chosen patterns are situated (Fig. 8). In the case of lesions, it is to be expected that analysis with patterns of larger bandwidth will lead to the same transfer function shown in Fig. 8, curve $\tilde{F}_2$.

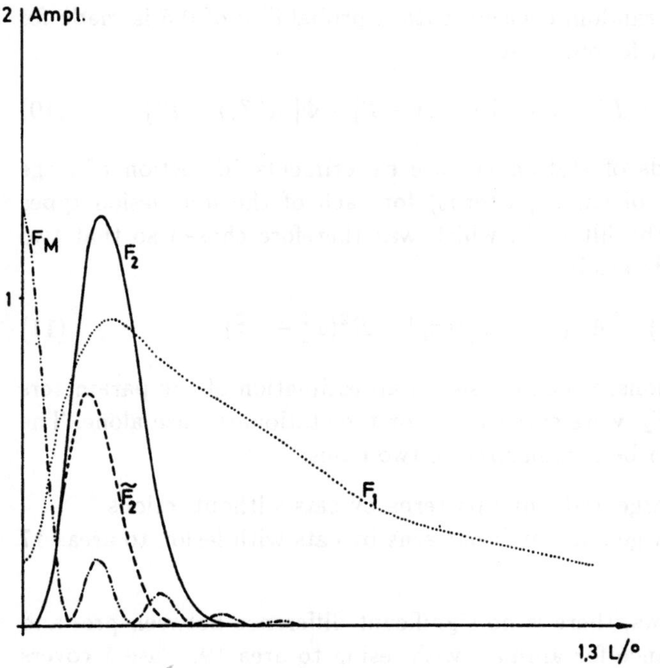

Fig. 8. Filter characteristics of the model and the signal. F_1 filter characteristic describing the retina; F_2 filter characteristic describing the normal state (see F_2 in Fig. 7); $\tilde{F}_2$ filter characteristic describing the lesion of areas 17 and 18; F_M frequency content of the signal (*circle*)

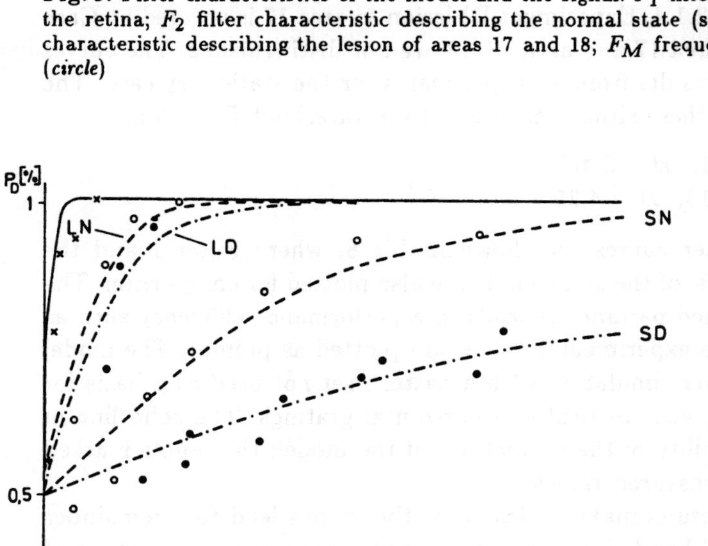

Fig. 9. Performance of the model. *Dashed lines* simulation of the above experiments. *LN* large patterns, normal state; *LD* large patterns, with lesion; *SN* small patterns, normal state; *SD* small patterns, with lesion. *Solid line* predicted discrimination of vertical and horizontal square wave gratings

2.3 Results

The procedure chosen to measure performance in the case of disturbance is sufficiently sensitive and suitable for system analysis when as many subsystem combinations are eliminated (lesioned) as the number of couplings which have to be determined. The model thus provides a hint as to the tasks performed by the different areas, assuming that the functional loss in the experiments is the *direct* result of the lesions and not a consequence of damage to auxilliary systems.

1. *Midbrain path*: The lesion of areas 17, 18, and 19 shows that the midbrain path (via the superior colliculus) contributes nothing without retraining.
2. *Area 19*: Lesion of areas 17 and 18 allows the determination of that part of performance taken over by area 19, as is shown in Fig. 4. Area 19 does not seem to participate in the task in the stationary case in the normal animal. On the other hand, when there is very slow movement, area 19 does seem to be involved (Fig. 5). The velocity range determined in the behavioral experiment coincides with that ascertained neurophysiologically in a substantial number of cells.
3. *Area 17*: Lesion of area 17 causes the same deficit in the stationary case as lesion of areas 17 and 18. Thus the difference between the curves F_2 and $\tilde{F}_2$ in Fig. 9 indicates the noncompensable deficit, and curve $\tilde{F}_2$ characterizes area 19. Because of the spectral content of the pattern, the high frequency range was not measured. One can, however, assume that area 18 overlaps this range.
4. *Area 18*: This area does not seem to be involved in the stationary case (see above), but it may come into play when the patterns move. The importance of area 18 in the case of a moving pattern is confirmed by neurophysiological findings (Dinse and von Seelen 1981).

In view of the number of patterns the cat is able to distinguish, it is doubtful whether the derived model is in fact implemented, especially when one considers the memory capacity involved in this classifier. On the other hand, other research groups have obtained similar results even in the analysis of human pattern recognition ability. We suspect that the reason for such findings is that the system studied is time variant and configures itself in dependence on the complexity level of the task to be performed. Simple discriminations always lead to filtering and cross correlation (Türke 1981). More complicated discriminations seem at first to cause more complex feature extraction, but do not alter the assignment operation, which is equivalent to the search process in associative memory. Such structures are, however, purely hypothetical. To obtain a better foundation for the ideas outlined here, we have attempted to model a part of the cortex. In doing so, we have proceeded on the hypothesis that such a system provides sufficiently flexible hardware for a synthesizing approach which would, for example, clarify extraction operations.

3 Dynamical Model Derived from Cortex Anatomy

Cortices are distinguished from other neuronal systems by some obvious structural principles, such as homogeneity in the tangential plane and inhomogeneity perpendicular to it. This section deals with a model of the cerebral cortex developed to study the functional consequences of cortical organization.

One of the structural principles is the lamination which we think to be important for several reasons: (1) Obviously, the different appearance of the layers indicates differences in their wiring. (2) The layers are distinguished by their connectivity. Specific thalamic afferents, for example, terminate mainly in layer IV, afferent fibers from cortical areas in the supragranular layers. The majority of cortico-cortical efferents stem from supragranular layers, whereas cortico-thalamic efferents originate from infragranular layers. (3) The response characteristics of a neuron depend to a large extent upon the layer it belongs to (cf. Gilbert 1977). We therefore consider a cortical area as a multi-input multi-output system composed of few functionally different layers.

Despite the multiplicity of shapes of cortical neurons, they can be classified into a few cell classes, the most frequent being the pyramidal cell. The dendritic trees of these neurons mostly reach from the layer where the cell body is situated up to layer I. This geometrical design suggests that information processing consists in mixing information from different layers in a systematic way: the deeper a neuron is located, the more layers provide it with input. In addition, the axonal arborization of the pyramidal cell extends into several layers, so that the output is also distributed to various layers. Since these intrinsic connections outnumber the contacts between sensory afferents and cortical neurons by some two orders of magnitude (Braitenberg 1981) neuronal information processing essentially means combining the instantaneous external input with a large amount of already processed information. The processing is predominantly performed by excitatory connections. Inhibitory influences are mediated by a second cell type, the nonspiny stellate cells, whose axonal and dendritic trees are situated in the proximity of the cell bodies.

To study the functional consequences of the principles indicated so far (i.e., lamination, intrinsic feedback, distinct preponderance of excitation), we transformed the anatomical description into a functional one by using the concept of "mean anatomy". We define an "average neuron" by the densities of its pre- and postsynaptic sites, or of its dendritic and axonal arborizations respectively. Their distributions are ascertained by observing a large number of neurons of the same type and then defining three-dimensional dendritic and axonal clouds. The usefulness of mean anatomy rests on two assumptions. First, we think that the coupling of cells is largely determined by the amount of overlap of their dendritic and axonal arborizations. Second, we claim that an average system exhibits average behavior, at least as long as its elements are linear.

We also take into consideration the temporal aspects of the propagation of neuronal activity, so that the temporal development of the spatial distribution of excitation can be studied. Because of the homogeneity and continuity of the layers we are able to describe the behavior of the system using the systems theory of homogeneous layers (von Seelen 1968, Marko 1969). In addition, most simulations presuppose linearity of the model. In contrast to the model proposed by Wilson and Cowan (1973), ours assumes several layers and stresses their functional differences.

3.1 Structure of the Model

The general anatomy of cortical areas was determined as outlined in the preceeding section. In the case presented in this paper, anatomic data were taken from the mouse (*Mus musculus*) cortex (Krone et al. 1986), but the model is in general equally well suited to deal with other species or different cortical areas.

We computed the space and time dependent responses of pyramidal cells in five two-dimensional cortical layers (II to VI) to input stimuli into layer IV. The calculation of neuronal coupling is based on the assumption that the degrees of overlap between "clouds" of pre- and postsynaptic sites (representing the axonal and dendritic aborizations of neurons) weighted with the densities of sites in these clouds can be used as a measure of the average strength of coupling between two neurons at a given distance from each other. Thus, a neuron is represented in the model as a set of (rotationally symmetric) density distributions of its pre- and postsynaptic sites in the six different layers.

The inhibitory connections are not modelled by separate inhibitory interneurons but simply by an additional distribution of inhibitory (i.e., negative) inputs to the excitatory neuron. They are given an additional synaptic delay, and, in the simulations, are restricted to the layer in which the neuron is situated, according to the observation that inhibitory synapses are found mostly at, or close to the cell body of the postsynaptic neuron. Compared to the separate modelling of inhibitory neurons, this means essentially only one restriction: there is no direct coupling in the model between inhibitory neurons.

Let us denote by $d_{ij}^{+}(x,y,x',y')$ the density at (x',y') in layer i of excitatory postsynaptic sites of a neuron at (x,y) in layer j (i.e., whose cell body is at (x,y) in layer j), by $d_{ij}^{-}(x,y,x',y')$ the corresponding density of inhibitory postsynaptic sites (in our model usually restricted to $i=j$), and by $a_{ji}(x,y,x',y')$ the corresponding density of presynaptic sites. Then we obtain the strength of coupling from a neuron at (x,y) in layer m to a neuron at (u,v) in layer n as

$$K_{mn}(x,y,u,v) = \sum_{j=1}^{6} \int\int a_{mj}(x,y,x',y') \times (d_{jn}^{+}(u,v,x',y') - d_{jn}^{-}(u,v,x',y'))dx'dy' \ . \qquad (12)$$

In fact, the functions d_{ij}^{+} and d_{ij}^{-} represent not only the densities of synaptic sites but also incorporate the strength of the effect that a synapse has on the postsynaptic cell. Usually, this strength is supposed to increase as the distance from the cell body decreases. The density of cells in the different layers and the thickness of the layers is incorporated as a factor in a_{ji}.

The cortical layers are modelled by means of the systems theory of homogeneous layers (von Seelen 1968, Marko 1969). Equation (12) is then reduced to a simple convolution since K_{mn} depends only on the difference of the spatial coordinates. In the Fourier domain, Eq. (12) becomes:

$$K_{mn}(\omega_1,\omega_2) = \sum_{j=1}^{6} a_{mj}(\omega_1,\omega_2)[d_{jn}^{+}(\omega_1,\omega_2) - d_{jn}^{-}(\omega_1,\omega_2)] \ . \tag{13}$$

Here, the symbols d^+, d^-, a and K are used for the Fourier transforms of the corresponding quantities in Eq. (12).

Equation (12) only describes the spatial characteristics of the model. The temporal characteristics of neurons are also represented in a very simple way. Our neurons are modelled as linear systems that basically add up their input and then produce their output with a certain temporal delay and a low-pass characteristic. In the simulations we use a discrete time, where one time step (0.1 ms) corresponds to the propagation time (axonal or dendritic) between one layer and the next. In this framework we can model the synaptic low-pass by the recursion $y(t) = x(t)/T + y(t-1)(1-1/T)$, with a time constant T of 5–10 ms. Synaptic delay is 1 ms. The resulting simulation procedure is illustrated in Fig. 10.

With these preliminaries, we can now develop the detailed model for the evolution in time t of the neuronal excitation $y_i(t;x,y)$ in layer i at (x,y) or its Fourier transform $y_i(t;\omega_1,\omega_2)$ with respect to the two spatial coordinates.

First we may form the matrices

$$A = A(\omega_1,\omega_2) := (a_{nj}(\omega_1,\omega_2))_{n,j} \ , \tag{14a}$$

$$D^{+-} = D^{+-}(\omega_1,\omega_2) := (d_{jn}(\omega_1,\omega_2))_{j,n} \tag{14b}$$

(here $n = 2,\dots,6$ and $j = 1,\dots,6$) and the vector $y = y(\omega_1,\omega_2) := (y_i(\omega_1,\omega_2))_{i=2,\dots,6}$. These matrices can be divided up into the submatrices $A(z)$, $z = 1,\dots,6$, where

$$a_{nj}(z) = \begin{cases} a_{nj}, & \text{if} \quad |n-j| = z-1 \\ 0 & \text{otherwise} \end{cases} \tag{15}$$

and similarly $D^+(z)$ and $D^-(z)$, namely into the parts with common delay z.

We denote by $x(t)$ the axonal input excitation, by $y(t)$ the output of the system (i.e., the excitation of the pyramidal cells), and by $e(t)$ the total axonal excitation. Then we can write the iteration as follows – always omitting the two frequency parameters:

$$e(t) = x(t) + \sum_{z=1}^{6} A(z)y(t-z) \ , \tag{16a}$$

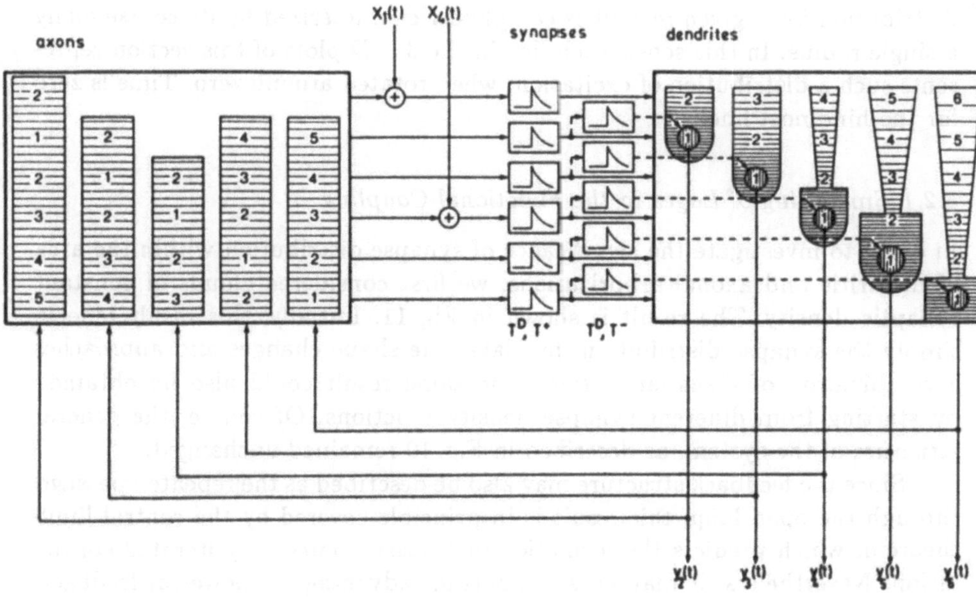

Fig. 10. Computational structure of the model. Let a small region of layer IV be stimulated by an input $x_4(t)$. After a time delay (T^D) and low-pass filtering (time constant T^+) at the excitatory synapses the cells sending dendrites to the stimulated region will be activated. In the next step the excitation $y_2, \ldots, y_6$ of the cells will spread over their axonal arborizations. The sum of activity in each layer together with the external input forms the new presynaptic activity and the loop starts again. The *dashed lines* represent the inhibition which is transferred by an additional synpase (T^- time constant of the low-pass filter), and acts on the cells within the vertically hatched region. The *numbers in the dendritic part* of the figure indicate the time delay after which postsynaptic activity present at a particular site is transformed into neuronal activation ($y_2, \ldots, y_6$). Analogously, the *numbers in the axonal part* indicate the propagation time on the axonal fibers

$$f^+(t) = e(t-10)/T^+ + f^+(t-1)(1-1/T^+) \tag{16b}$$

$$f^-(t) = f^+(1-10)/T^- + f^-(t-1)(1-1/T^-) \tag{16c}$$

$$y(t) = \sum_{z=1}^{6} (D^+(z)f^+(t-z) - D^-(z)f^-(t-z)) \; . \tag{16d}$$

Equations (16b,c) model synaptic delays and low passes and Eqs. (16a,d) represent spatial filtering and the propagation times.

3.2 Performance of the Model

Starting out from the spatiotemporal coupling of the individual layers, we computed the distribution of excitation in the entire feedback system, when externally stimulated by a δ-pulse in space and time. Since this distribution is a "landscape" developing in time, it usually cannot be depicted in a single plot. However, as all occurring transfer functions are rotationally symmetric, the distribution for a given instant is completely characterized by its course along a single radius. In this sense, each line in the $3-D$ plots of this section represents such a distribution of excitation, when rotated around zero. Time is zero for the hindmost line.

3.2.1 Smoothing of Edges in the Functional Coupling

In order to investigate the significance of synapse distribution within the area of dendritic and axonal arborizations, we first considered clouds of constant synaptic density. The result is shown in Fig. 11. Initially, this result reflects simply the synapse distribution, but later the shape changes and approaches a combination of Gaussian curves. The same result could also be obtained by starting from different synapse density functions. Of course, the general structure of the system, as described in Fig. 10 remained unchanged.

Since the feedback structure may also be described as the repeated passage through the open loop, this result is in principle covered by the central limit theorem, which predicts the formation of Gaussian curves by iterated convolution. Nevertheless, it may be an important advantage of powerful feedback that the functional coupling and consequently the mode of processing becomes independent from detailed wiring.

In what follows, we use Gaussian synapse density functions throughout.

3.2.2 Space-Time Interdependence

All simulations showed more or less conspicuous oscillations irrespective of the parameters used. The cycle durations were in the order of 50 ms (Figs. 12, 13).

Temporal behavior at a given point depends strongly on the point's distance from the input site. This is due to the relative strength of feedback versus direct coupling which have different time behavior. Obviously direct coupling is limited by the length of the nerve fibers, whereas feedback can reach to more

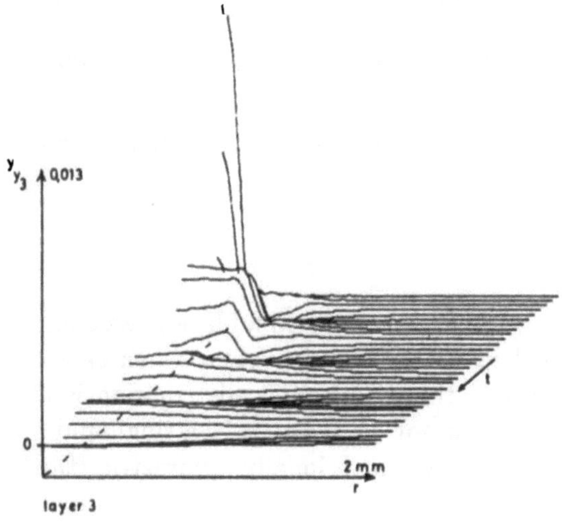

Fig. 11. Temporal development of functional coupling assuming clouds of constant synaptic densities within the dendritic and axonal arborizations. This figure shows the spatio-temporal distribution of activation in layer 3 in response to an impulse stimulation in layer 4 at time 0. Initially, the result reflects the synaptic density function, but later the edge is smoothed. The final result appears to be independent of the density function used (compare Fig. 13, layer 3). Conventions and parameters as in Fig.13

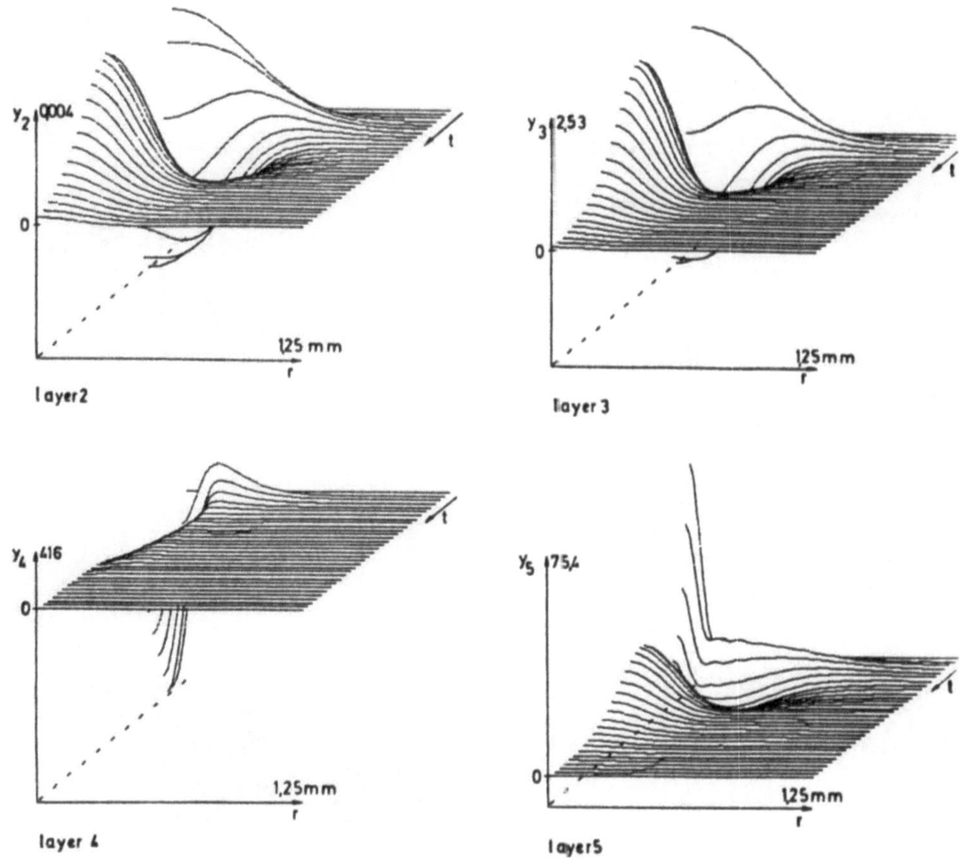

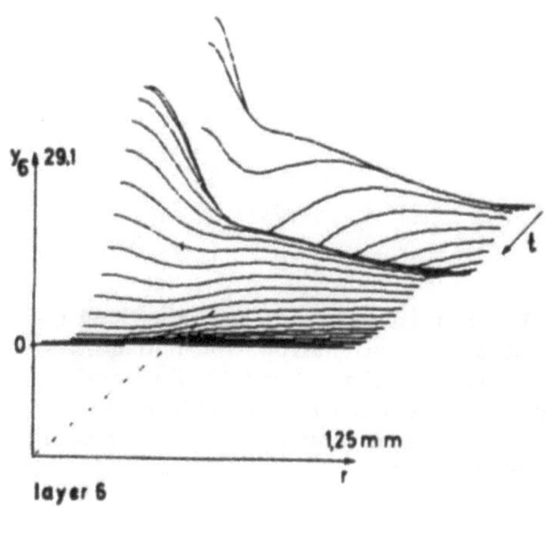

Fig. 12. Simulation results assuming relatively wide apical dendritic trees and few recurrent axons. At time zero (*hindmost line*) *layer 4* is stimulated by a spatiotemporal δ-impulse. Note the relatively simple results in *layer 4* and the different types of temporal oscillations in *2* and *3* as compared to *5* and *4*. Time varies from 0–100 ms, activation in arbitrary units

distant points. Therefore, the impulse response of the entire system cannot be separated into two mutually independent factors, describing spatial and temporal behavior respectively. As a consequence, the temporal integration of the impulse response may give rise to an entirely new functional coupling, whereas in the case of separability the spatial part would simply be reproduced. The

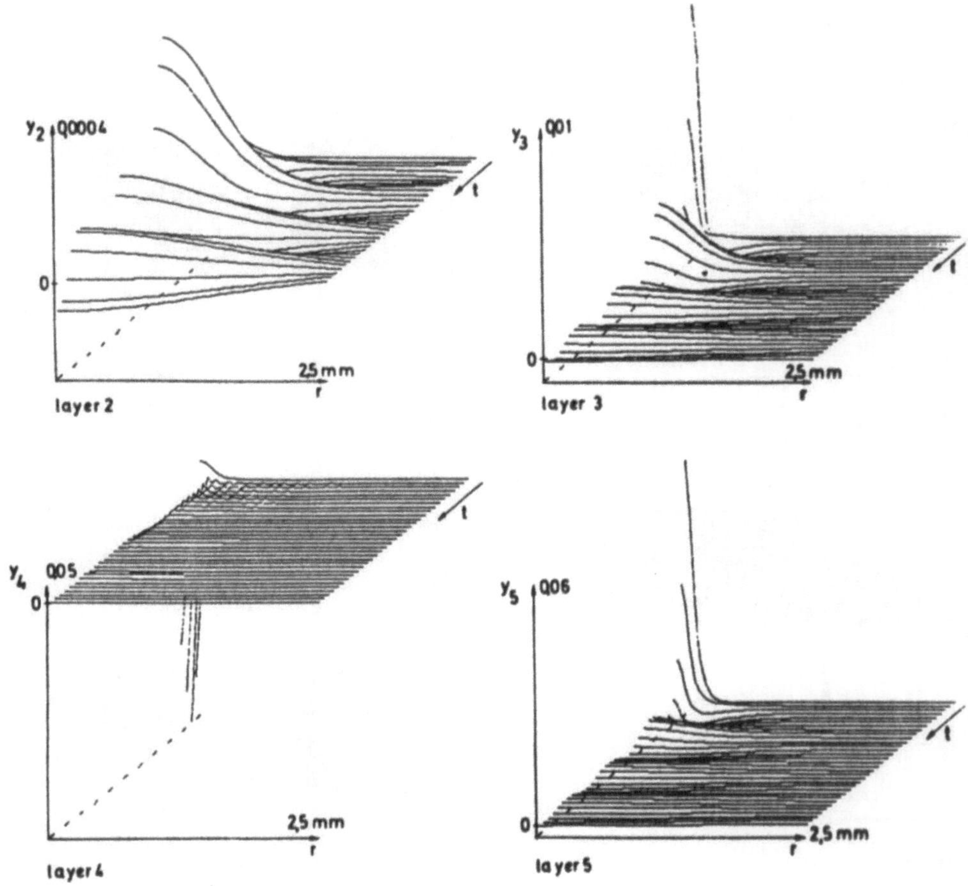

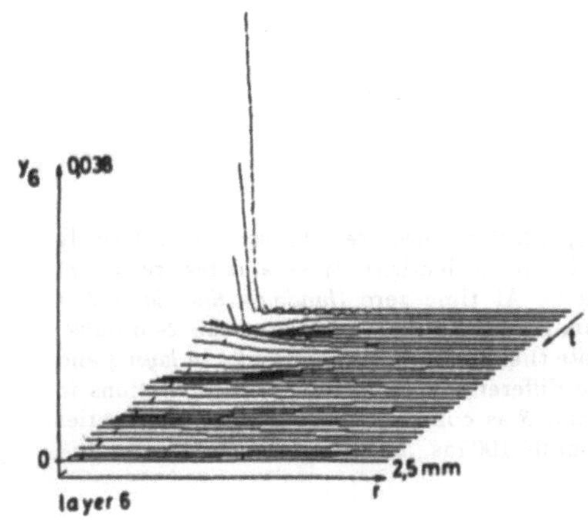

Fig. 13. Simulation results assuming long recurrent axons and narrow apical dendrites. Same input as in Figs. 11 and 12. Temporal oscillations are clearly visible in all layers except *layer 4*. Time varies from 0–150 ms, activation in arbitrary units

above result implies that a sustained stimulus will be processed by a spatial filter which is essentially different from the filter applied to a transient stimulus.

Figure 14 depicts the responses to a spot of sustained stimulation (step response) computed from the impulse responses shown in Fig. 12. When a steady state is reached, lateral bands (i.e. rings) are clearly visible, in the cases in

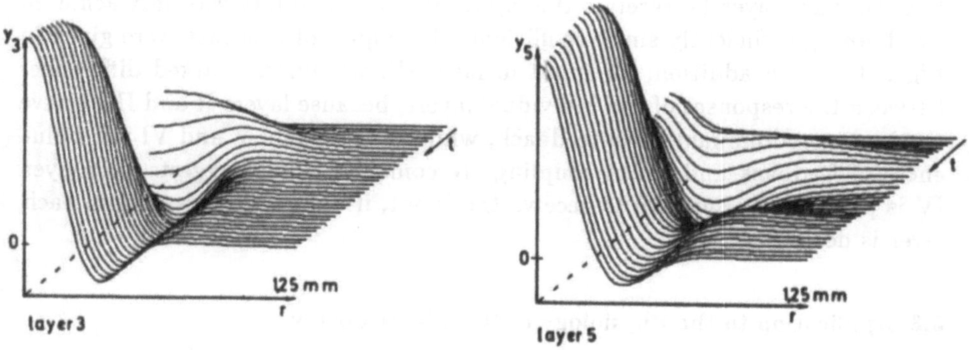

Fig. 14. Responses to a temporal step function applied to *layer 4* assuming the same parameters as in Fig. 12, *layers 3* and *6* are shown. Summation over time of the lines in Fig. 12 leads to a new functional coupling which is seen in the foremost line (steady state). Thus a functional equivalent to lateral inhibition is generated although the range of inhibitory neurons is assumed to be shorter than that of excitatory ones. Same conventions as in Fig. 12

Fig. 12 in the form of lateral inhibition. This lateral inhibition arises in the functional coupling as an effect of the feedback structure although, in our model, inhibitory neurons have shorter fibers than excitatory ones.

In a nonlinear modification of the model, where negative excitations were cut off, similar lateral bands were obtained by sustained stimulation. An example is shown in Fig. 15.

3.2.3 Specificity of the Layers

The simulated excitation distributions differ systematically for each layer and also depend on which layer receives the input. For instance, when the stimulus is applied to layer I, the excitation disperses and finally vanishes without oscillations or any other interesting development. It is remarkable that this does not

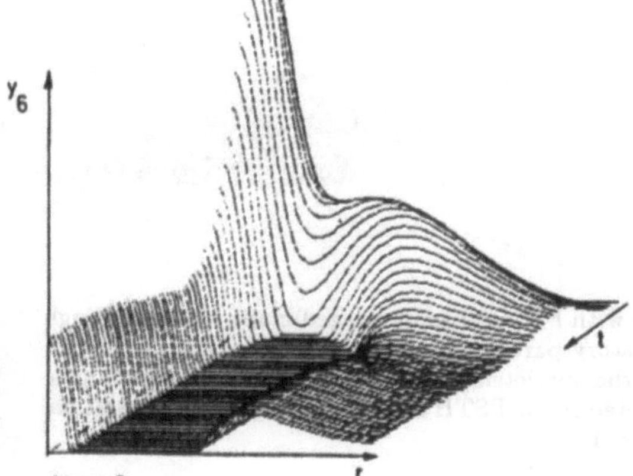

Fig. 15. Temporal step responses of a nonlinear modification of the model, where negative values were cut off at each time-step. Note the generation of lateral bands comparable with those depicted in the linear simulation, Fig. 14. The sustained stimulus (step function) was applied to *layer 4*; the reaction of *layer 6* is shown. Same parameters and conventions as in Fig. 12

happen when layer IV receives the input, although stability was only achieved by choosing sufficiently small coefficients. Examples of this case were given in Figs. 11–15. In addition, for input in layer IV, we obtain marked differences between the responses of the individual layers, because layers II and III receive the input predominantly via feedback, while in the layers V and VI the influences of feedback and direct coupling are combined. This suggests that layer IV is particularly suitable to receive the input, if different processing in each layer is desired.

3.3 Application to the Physiology of the Visual Cortex

In this section we discuss the model and the simulation results in the context of physiological experiments. For such a discussion we should keep in mind that the model cannot represent the individual behavior of single neurons but rather the "climate" of basic general properties of areas, in which the individual neurons can develop their individuality.

In a linear, space-invariant model, the spatial impulse response may be identified with the weighting function of the "receptive field" (RF) of any unit in the neural layer. Similarly, the temporal behavior of the impulse response is a model of the post stimulus time histogram, PSTH. An example is given in Fig. 16.

Oscillations with a period in the order of 50 ms occur frequently in electrophysiological recordings (Fig. 16) and in registrations of visually evoked potentials. In our model, such oscillations were most markedly obtained in layers III and V. In the case of an oscillating response, one will usually not succeed in

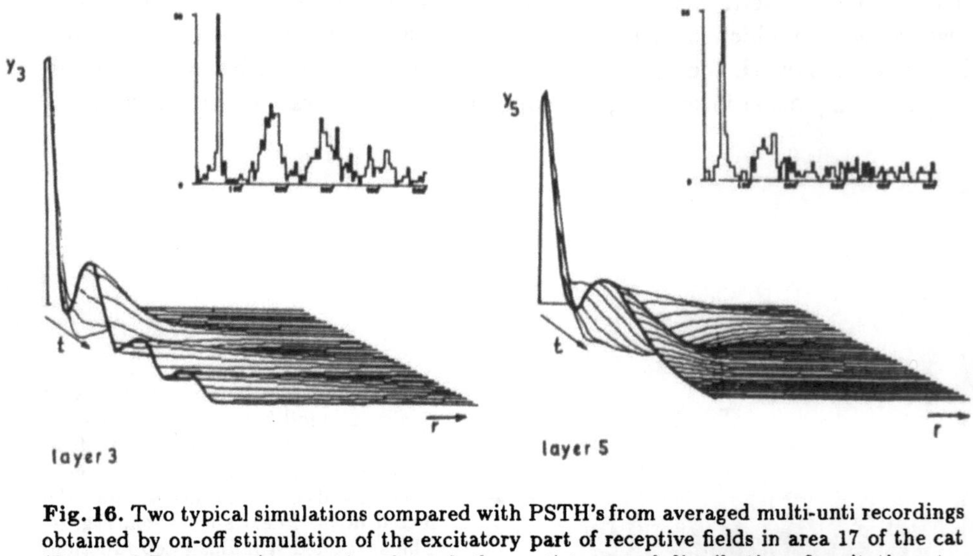

Fig. 16. Two typical simulations compared with PSTH's from averaged multi-unti recordings obtained by on-off stimulation of the excitatory part of receptive fields in area 17 of the cat (Best and Dinse 1984). A section through the spatiotemporal distribution of excitation at a fixed coordinate (e.g., zero) may be interpreted as the PSTH of a neuron. *Left* same simulation as in Fig. 13; *right* same simulation as in Fig. 12

dividing a field into excitatory and inhibitory subfields. It is, therefore, quite possible that units behaving in the way predicted for layers III and V will be classified as complex cells by the electrophysiologist. Complex cells are indeed said to occur predominantly in layers III and V (Gilbert 1977).

The latencies for on-off stimulation of the excitatory receptive field are shortest in layer IV and increase for each following layer by up to 20 ms (Fig. 17; Best and Dinse 1984). Clearly, this cannot be an effect of synaptic delay and propagation time alone as these would account for only a few milliseconds at the most. According to our model, the excitation passes several feedback loops during this time (cf. Fig. 12, layer VI, and Sect. 3.2.1). In this sense, the time is required to build up the receptive field.

As discussed previously, the temporal behavior of our simulations is usually different at every point in the field. Vice versa, the spatial coupling is different at every instant. Therefore, when the response to a light bar is simulated, the preferred length of such bars will usually also vary with time. This predicts that neurons will be tuned to different bar lengths at different times, i.e., the tuning curves obtained from different peaks of the PSTH will not be equal. A similar effect has in fact been found by Brönnimann and Dinse (1981) (cf. Fig. 17) and by Nothdurft (1983): bar length tuning for moving stimuli varies greatly with the velocity of movement.

It has been shown that detailed measuring of the RF function may establish a spatial shift of the excitatory region in the course of time (Fig. 18; Krause and Eckhorn 1983, Dinse and Best 1984). Non-separability of spatial and temporal behavior is required to account for this effect. In principle, our model produces a similar effect, but, since it is restricted to rotationally symmetric functions, it exhibits e.g., a narrowing of concentric rings rather than a real shift.

Thus it should be clear that internal feedback plays a basic role in determining the general response properties of neurons. The most interesting feature introduced by this feedback is the complicated temporal behavior of the RF function. Functionally, this means that the spatial filter applied to a given stimulus depends on the temporal structure of that stimulus. Theoretically, we consider the temporal behavior to be a fruitful extension of the receptive field concept. With this concept one can, for instance, explain some properties of receptive fields which were hitherto supposed to required nonlinear treatment.

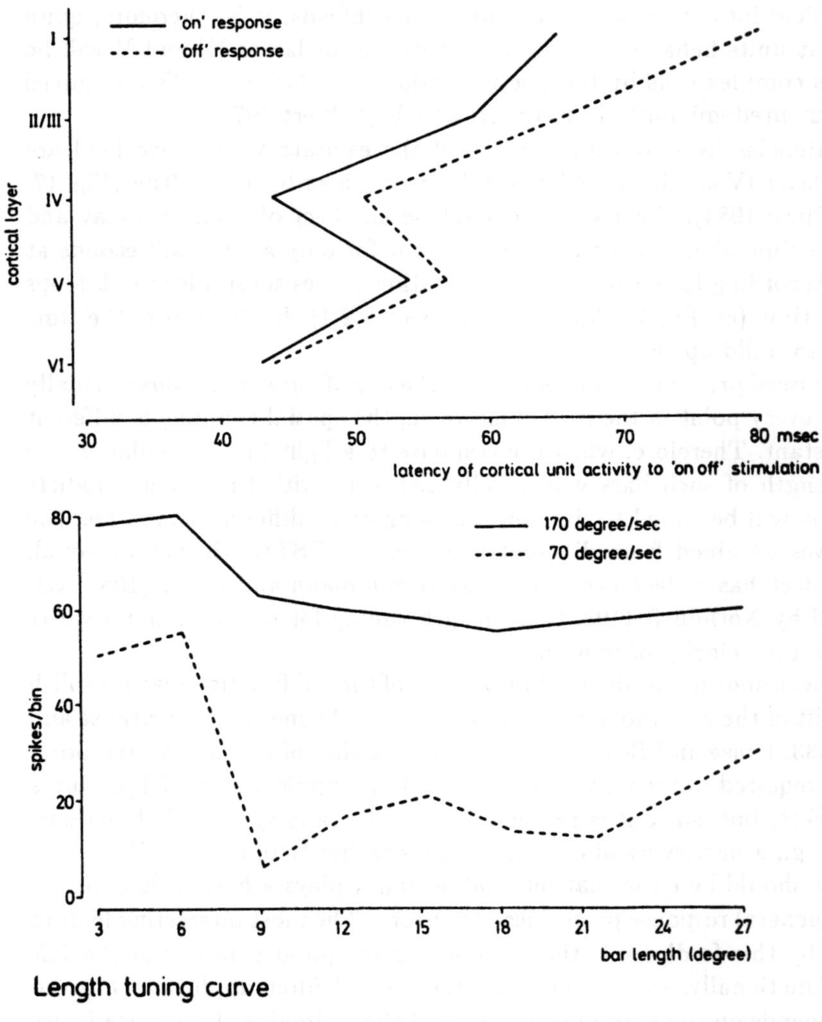

Fig. 17. *Top* Distribution of averaged latencies within cortical layers for onset of cortical unit activity following on-off photic stimulation (sample includes *C* cells only). The short latencies in layer *IV* and *VI* suggest that these layers receive input. The long transmission time between the layers is probably processing time (positive feedback), rather than synaptic delay and conduction time which could account for a few ms only, but not for the differences in the range of up to 20 ms measured. *Bottom* Response of an Area 18 hypercomplex cell stronlgy inhibited when stimulated with elongated bars (*dashed line*). When the velocity of the stimulus is increased (*solid line*), the pronounced end zone inhibition vanishes almost entirely and the remaining small decrease in activity can no longer supply a sufficient argument for classifying this cell as hypercomplex. This experiment provides further evidence that distribution of inhibition perpendicular to the axis of movement is controlled to a large extent by temporal parameters and is thus not exclusively a spatial property of the *RF* organization

Fig. 18. Space-time-diagram of an Area 17 *C* cell. *RF* was devided into subfields each of which was scanned with a small spot. Based on the resulting response planes, the pattern of temporal development of excitation across the grid of the subfields was calculated using 20 ms steps for temporal resolution. The complete procedure resulted in "films" of 200-ms duration, each single picture of which displays the current spatial distribution of excitation

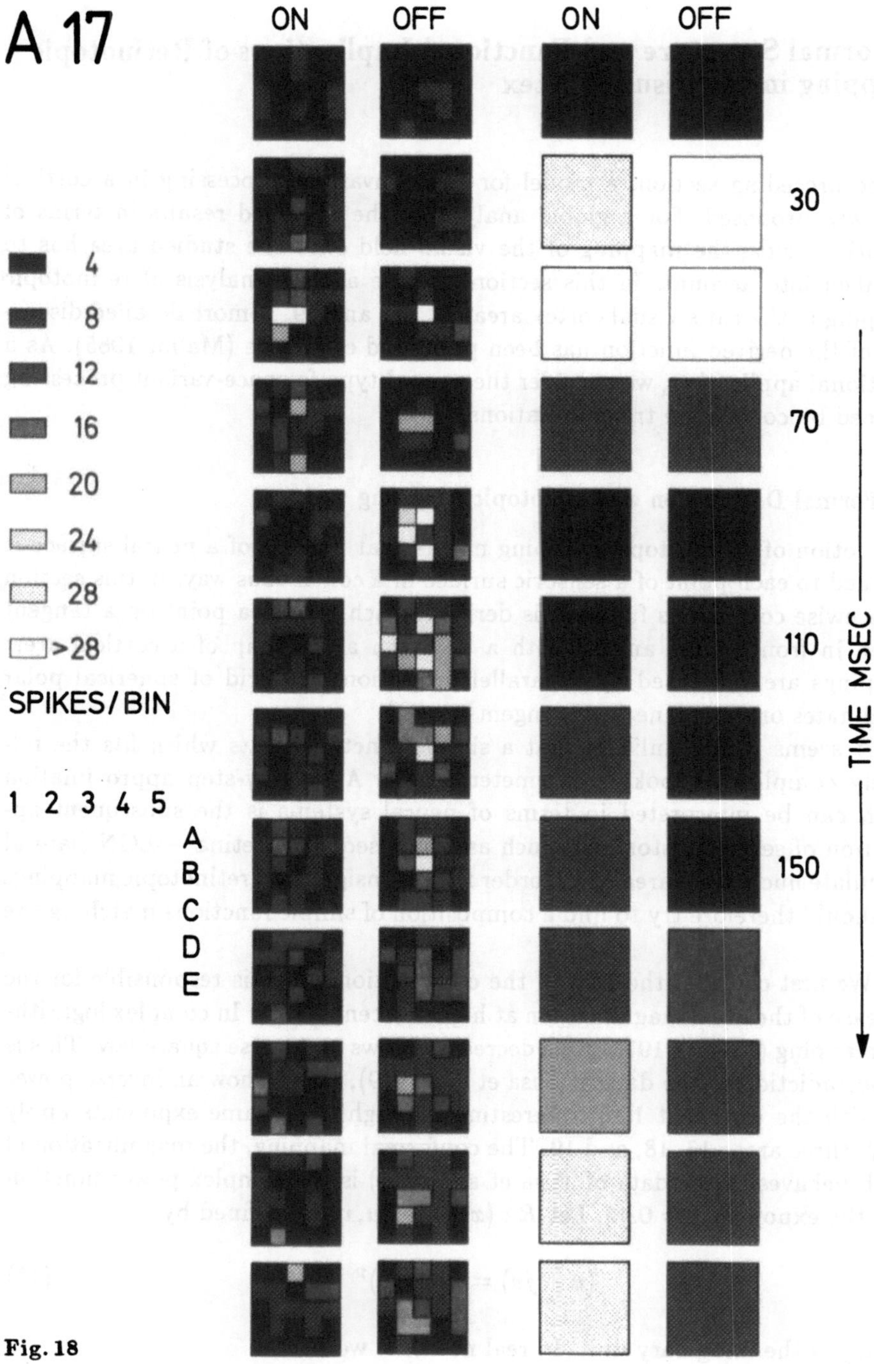

Fig. 18

within the entire *RF*. On the *right* are films obtained following whole field stimulation. First, the oscillation of excitation is not spatially homogeneous, but rather every position within the *RF* exhibits its own temporal characteristic, which, in the film, gives the excitation the appearance of a wave spreading over the entire *RF*. Second, when these results are contrasted with whole field stimulation (*left side*), striking non-linearities become evident, suggesting strong time-dependent summation and reciprocal inhibition between different positions of the *RF*

4 Formal Structure and Functional Implications of Retinotopic Mapping in the Visual Cortex

In the preceding section, a model for space-invariant processing in a cortical area was proposed. For a global analysis of the obtained results in terms of receptive fields, the mapping of the visual field onto the studied area has to be taken into account. In this section we give a brief analysis of retinotopic mapping in the cat's visual cortex areas 17, 18, and 19. A more detailed discussion of the derived function has been published elsewhere (Mallot 1985). As a functional application, we consider the special type fo space-variant processing induced by coordinate transformations.

4.1 Formal Description of Retinotopic Mapping

The notion of receptotopic mapping means that a point of a neural surface is assigned to each point of a sensoric surface in a continuous way. In this section a piecewise continuous function is derived which relates a point on a tangent screen in front of the animal with a point on a flat map of a cortical area. Mappings are visualized by a parallel projection of a grid of spherical polar coordinates onto a plane (i.e., tangent screen).

It seems rather unlikely that a simple function exists which fits the relatively.complicated looking perimeter charts. A step-by-step approximation which can be interpreted in terms of neural systems is the subsequent application of several distortions, such as in the sequence retina – LGN (lateral geniculate nucleus) – area 17. In order to gain insight into retinotopic mappings one should therefore try to find a composition of simple functions matching the data.

We first consider the part of the composition which is responsible for the decrease of the areal magnification at higher eccentricities. In complex logarithmic mapping (Fischer 1973), this decrease follows an inverse square law. This is in contradiction to the data of Tusa et al. (1979), which show an inverse power law with the exponent 1.13. Interestingly, roughly the same exponents apply for all three areas 17, 18, and 19. The conformal mapping, the magnification of which behaves as the data of Tusa et al. (1979) is the complex power function with the exponent $p = 0.43$. Let $R : (x, y) \rightarrow (u, v)$ be defined by

$$(u + jv) = (x + jy)^p \ , \tag{17}$$

where j is the imaginary unit. In real notation we have

$$R : (x, y) \rightarrow (u, v) := (x^2 + y^2)^{p/2} \begin{pmatrix} \cos [p(2n\pi + \arctan y/x)] \\ \sin [p(2n\pi + \arctan y/x)] \end{pmatrix} \tag{18}$$

where n is a suitable integer.

The main difference between the maps of areas 18 and 19 and that of area 17 is that the outer part of the horizontal meridian is represented twice in the areas 18 and 19. In the complex power function with an exponent less than unity, this behavior is well known for the left (i.e., negative) part of the real axis. It is due to the ambiguity of the arctan function (Eq. 18). If the perimeter charts of areas 18 and 19 on the one hand, and area 17 on the other are to be described by one function, it would thus be sufficient to mirror the visual hemified in order to obtain negative coordinates in the precursor of the maps of areas 18 and 19.

Figure 19b shows the transformation of the entire visual field by the complex power function ($p = 0.43$). The upper and lower left side of the plot may be regarded as a first approximation of the area 18 map, bordering on area 17 (right) along the vertical meridian. Figure 19c shows the function $(u, v) = R[(x, y) + (1, 0)]$ (cf. Eq. 18), i.e. an eccentric power function. Here, the horizontal meridian in the left part is divided for larger eccentricities only. However, this leads to a new problem: in the area 18 map in Fig. 19c, magnification is largest not at the area centralis, but at the branch point, where the horizon-

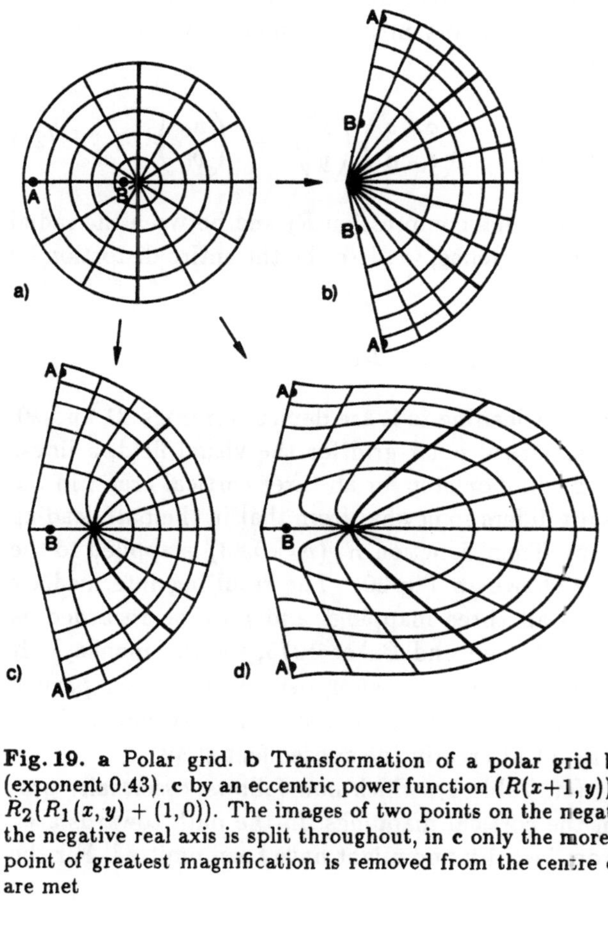

Fig. 19. **a** Polar grid. **b** Transformation of a polar grid by the complex power function (exponent 0.43). **c** by an eccentric power function ($R(x+1, y)$). **d** by the two-step-modification $R_2(R_1(x, y) + (1, 0))$. The images of two points on the negative real axis are indicated. In **b** the negative real axis is split throughout, in **c** only the more peripheral part is split, but the point of greatest magnification is removed from the centre of gaze. In **d** both requirements are met

tal meridian forks. It is therefore necessary to separate the part of the function responsible for the magnificaion from that generating the discontinuity.

This is done by decomposing the function R as defined in Eq. 18 into two parts, R_1 and R_2, $R = R_2 \circ R_1$, where "$\circ$" stands for composition. R_1 contains the radial compression responsible for the magnification factor:

$$R_1 : (x,y) \rightarrow (s,t) = (x,y) \cdot (x^2 + y^2)^{(p-1)/2} \ , \tag{19}$$

R_2 contains the tangential compression, which gives rise to the discontinuity:

$$R_2 : (s,t) \rightarrow (u,v) = (s^2 + t^2)^{1/2} \begin{pmatrix} \cos\left[p(2n\pi + \arctan t/s)\right] \\ \sin\left[p(2n\pi + \arctan t/s)\right] \end{pmatrix} \ . \tag{20}$$

R_2 is equal-area except for a constant $2p$.

Figure 19d shows the transformation of the polar grid obtained by the function $(u,v) = R_2[R_1(x,y) + (1,0)]$, i.e., the shift necessary to keep part of the negative axis undivided is introduced only after the radial compression. Thus, the point of largest magnification has become independent from the branch point. Clearly, the mapping in Fig. 19d is no longer conformal.

Next, we consider additional equal-area distortions. For the sake of simplicity, we use only shifts and linear distortions, i.e., compositions of rotations, magnifications and mirroring:

$$L : \begin{pmatrix} x \\ y \end{pmatrix} \rightarrow \begin{pmatrix} u \\ v \end{pmatrix} := \begin{pmatrix} a & b \\ c & d \end{pmatrix} \begin{pmatrix} x \\ y \end{pmatrix} + \begin{pmatrix} u_0 \\ v_0 \end{pmatrix} \tag{21}$$

They are introduced prior to the radial compression R_1 and between the radial and the tangential compression R_2. Hence, we describe the entire distortion by a composition of four steps:

$$R = R_2 \circ L_2 \circ R_1 \circ L_1 \ . \tag{22}$$

The final model and the steps leading to it are depicted in Figs. 21 and 20, respectively. Starting with a spherical polar grid for the visual field, a linear distortion L_1 is performed as a common step for all three cortical areas. In the path leading to area 17 the shift differs somewhat from that in the path leading to areas 18 and 19 (cf. Table 2). Then, function R_1 ($p = 0.43$) is applied to the two shifted grids (only one is shown in Fig. 20). The resulting intermediate stage is essentially common to all three mappings and may be regarded as a model of the LGN map (cf. Malpeli and Baker 1975, for the monkey). It exhibits the difference in total area between the upper and the lower part of the visual hemifield. The decrease in magnification at greater eccentricities is already fully developed and is not altered in the subsequent steps.

The linear distortion L_2 (cf. Table 2), which now follows, essentially produces all differences between the cortical mappings. In the path towards area 17, it deviates only a little from identity and might have been omitted. For the

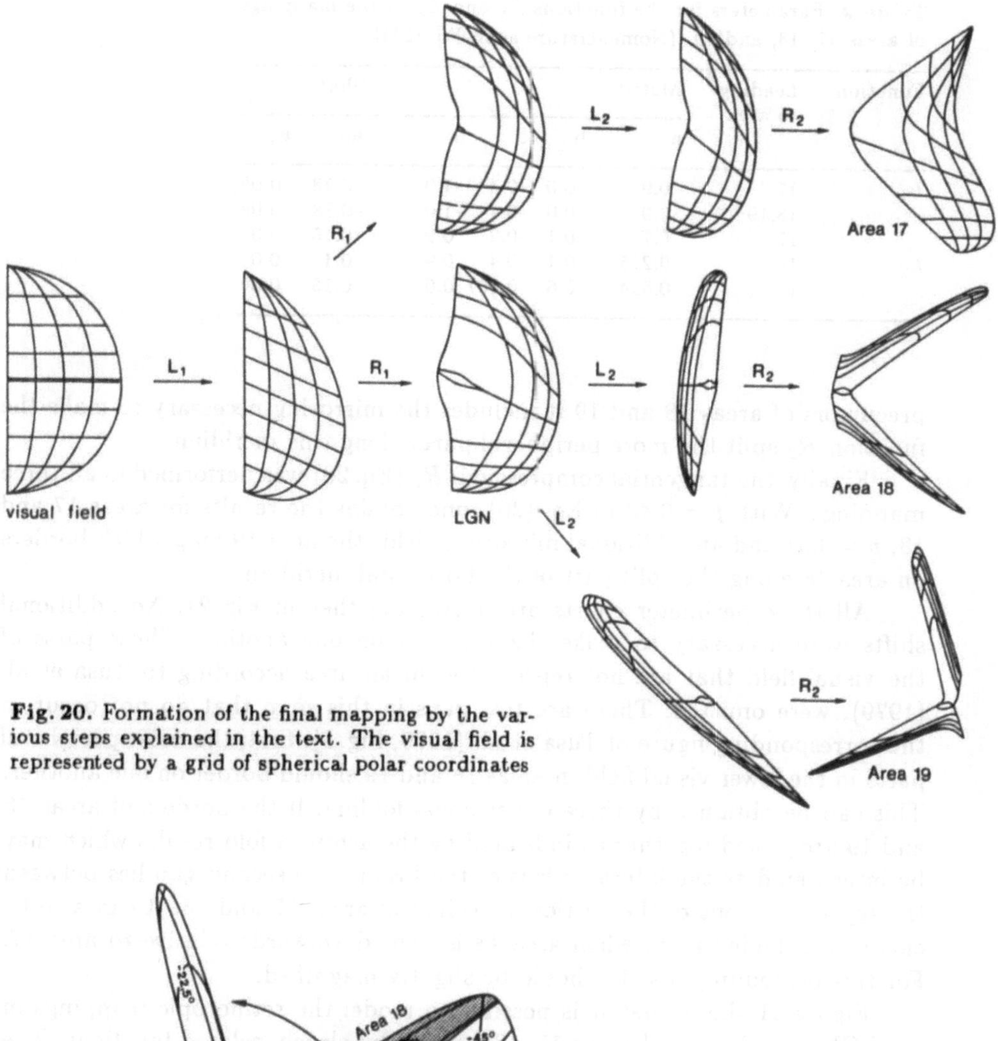

Fig. 20. Formation of the final mapping by the various steps explained in the text. The visual field is represented by a grid of spherical polar coordinates

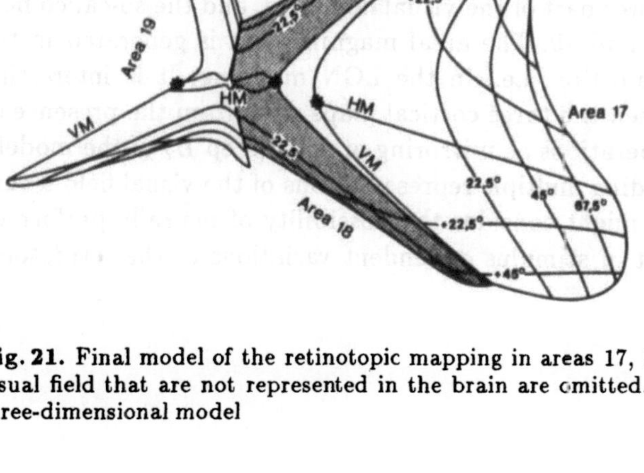

Fig. 21. Final model of the retinotopic mapping in areas 17, 18 and 19. Those parts of the visual field that are not represented in the brain are omitted. The gaps can be closed in a three-dimensional model

Table 2. Parameters for the functions L_1 and L_2 in the mappings of areas 17, 18, and 19 [(Nomenclature as in Eq. (21)]

Function	Leading to area	Matrix				Shift	
		a	b	c	d	u_0	v_0
L_1	17	0.9	0.0	−0.4	−1.0	0.18	−0.08
	18,19	0.9	0.0	−0.4	−1.0	−0.18	0.08
	17	0.7	0.1	−0.2	0.9	0.15	0.0
L_2	18	−0.225	0.1	0.4	0.9	0.15	0.0
	19	−0.534	−0.6	0.267	0.6	0.15	0.0

precursors of areas 18 and 19 it includes the mirroring necessary to make the function R_2 split the more peripheral parts along this meridian.

Finally, the tangential compression, R_2 (Eq. 20) was performed in all three mappings. With $p = 0.43$ in Eq. (20), one obtains the results for areas 17 and 18; $p = 0.57$ and an additional mirroring yields the area 19 map which borders on area 18 along the split part of the horizontal meridian.

All three perimeter charts are drawn together in Fig. 21. No additional shifts were necessary to make them border on one another. Those parts of the visual field that are not represented in an area according to Tusa et al. (1979), were omitted. There are two gaps in this map that do not occur in the corresponding figure of Tusa et al. (1979, Fig. 2): first, the most peripheral parts in the lower visual field in areas 18 and 19 should border on one another. This can be obtained by three-dimensional folding. If the borders of areas 18 and 19 are joined together as indicated by the arrow, a fold results which may be interpreted as the lateral sulcus of the brain. The second gap lies between the representations of the vertical meridian in areas 17 and 18. It can also be closed in a folded map, when area 18 is bent downwards relative to area 17. For this procedure, area 17 should be slightly magnified.

Figure 21 shows that it is possible to model the retinotopic mappings in the LGN and the visual areas 17, 18, and 19 by closely related functions. The comprehensive model exhibits the general form of the mappings, the differences between the upper and lower part of the visual hemifield, and the so-called field discontinuity in areas 18 and 19. The areal magnification is generated in the first two steps of the distortion, i.e., in the LGN mapping. It is interesting that the differences between the three cortical maps stem from the presence or absence of such simple operations as mirroring or shift (step L_2 of the model), rendering the task of building multiple representations of the visual field a very simple one. Further, one might consider the possibility of neurally performed shifts, which would result in stimulus dependent variations of the retinotopic map.

4.2 Functional Relevance of Mapping

An impression of how mapping acts in the cat's visual cortex can be obtained from Fig. 22: A typical scene from the cat's environment and its cortical representations are shown. As a step towards a more system theoretic analysis of the composition of various mappings with an intracortical operation, we consider the space-variant linear operation which results from space-invariant filtering of a distorted image. In technical applications, the composition of mapping with space-invariant filtering has been used for the restoration of motion-degraded images (Sawchuk 1974). In order to characterize the resulting operator, we simulate the weighting functions of the corresponding receptive fields.

Suppose the processing in a cortical area can be described by a spatial convolution with a coupling function. We substitute the retinal coordinates for the cortical ones and obtain the weighting function as the kernel of the resulting integral equation. This kernel differs from the back-projected coupling function by the Jacobian of the retinotopic mapping.

Figure 23 shows contour lines of a cortical coupling function (Fig. 23b) and of the associated weighting function of the receptive fields (Fig. 23a). The cortical coupling function chosen is rotationally symmetric, but the resulting receptive fields are clearly oriented and direction specific. This example matches well with the data of Payne and Berman (1983), who measured the relation between receptive field properties and their absolute position in the visual field. They found that orientations at right angles to the direction towards the area

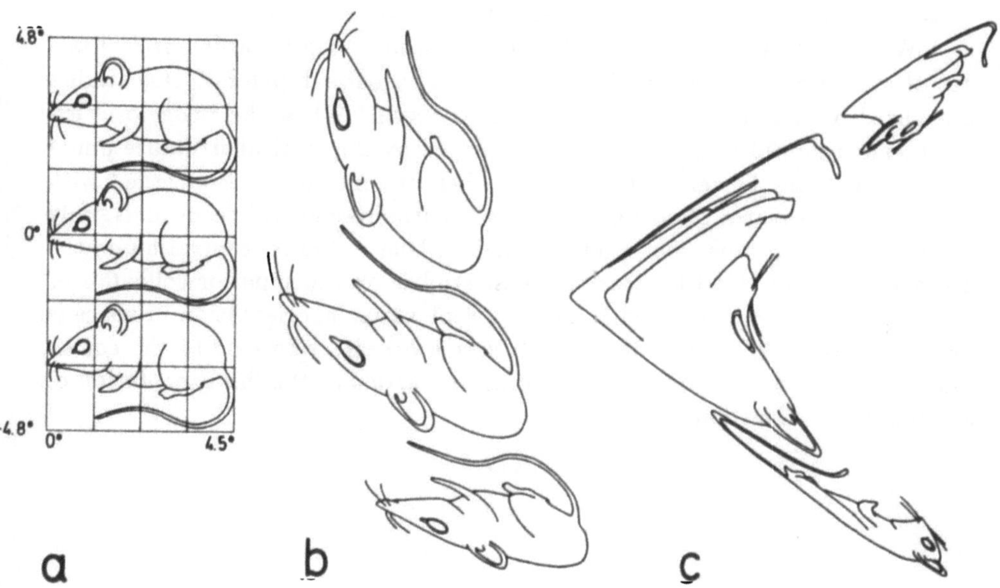

Fig. 22. **a** Scene in the central visual field. **b** Transformation according to the simulated area 17 map. **c** Transformation according to the simulated area 18 map

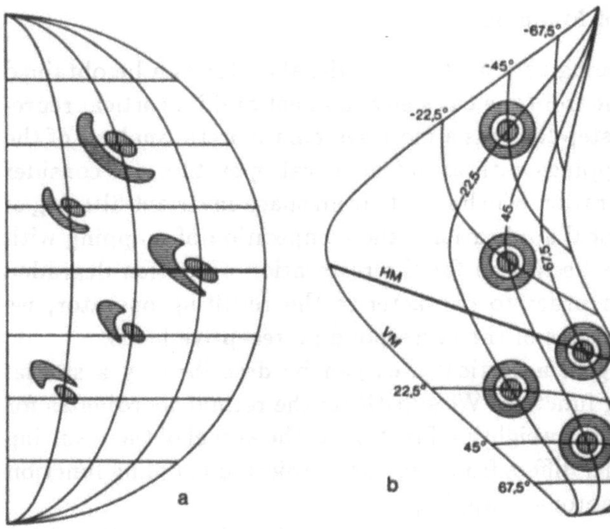

Fig. 23. b Contour lines of a cortical coupling function. a The associated weighting function of the receptive field. Intracortical coupling is symbolized by the difference of Gaussian functions. Although cortical coupling is isotropic, a given cell can be influenced most strongly from an asymmetric domain in the visual field. *Vertically hatched*: more than 30% of maximum excitatory influence; *horizontally hatched*: more than 70% of maximum inhibitory influence

centralis are more frequent than oblique ones.

In general, this analysis shows that a simple distortion of an image is probably not the only task performed via retinotopic mapping. Rather, it is possibly a powerful tool for space-variant image restoration.

We regard the three proposed models as a first step towards our real target of analyzing the cooperation between the visual subsystems. Our studies are based on the assumptions discussed in Sect. 1. While the behavioral experiments are used to quantify the part played by the cortical structure under consideration in the relevant task, the other two models are concerned with fundamental features of cortical systems. Neurophysiological recordings, only briefly mentioned here, round off the studies. Due to the concentration on the areas 17, 18, and 19 with occasional case studies on the superior colliculus or PMLS, the problems dealt with must be kept variable, orientation in space is also to be taken into consideration. It seems necessary to extend the cortex model to cover nonlinear system features, in particular the implementation of a comparison operation.

References

Best J, Dinse HRO (1984) Laminar dependent visual information processing in the cat's area 17. Neurosci Lett Suppl 18:S76

Braitenberg V (1981) Anatomical basis for divergence, convergence, and integration in the cerebral cortex. In: Grastyán E, Molnár P (eds) Adv Physiol Sci, vol 16. Pergamon Press, Oxford New York, pp 411–419

Brönnimann R, Dinse HRO (1981) Effects of velocity, direction and intensity on hypercomplex cells in the cat's area 18. Neurosci Lett Suppl 7:S147

Campbell FW, Maffei L, Piccolino M (1973) The contrast sensitivity of the cat. J Physiol (London) 229:719–731

Dinse HRO, Best J (1984) Receptive field organization of the cat's visual cortex exhibit strong spatiotemporal interaction. Neurosci Lett Suppl 18:S75

Dinse HRO, Seelen W von (1981) On the function of cell systems in area 18, Parts I and II. Biol Cybern 41:47–69

Fischer B (1973) Overlap of receptive field centers and representation of the visual field in the cat's optic tract. Vision Res 13:2113–2120

Gilbert CD (1977) Laminar differences in receptive field properties of cells in cat primary visual cortex. J Physiol (London) 268:391–421

Krause F, Eckhorn R (1983) Receptive fields for motion stimuli for different types of cat visual neurons. Neurosci Lett Suppl 14:S209

Krone G, Kunz D, Seelen W von (1983) On the analysis of the cat's pattern recognition system. Biol Cybern 48:115–124

Krone G, Mallot HA, Schüz A, Palm G (1986) Spatiotemporal receptive fields: A dynamical model derived from cortical architectonics. Proc Roy Soc Lond B (in press)

Krüger K, Heitländer-Fansa H, Dinse HRO, Berlucchi G (1986) Detection performance of cats lacking areas 17 and 18: A behavioral approach to analyse pattern recognition deficits. Exp Brain Res (in press)

Mallot HA (1985) An overall description of retinotopic mapping in the cat's visual cortex areas 17, 18, and 19. Biol Cybern 52:45–51

Malpeli JG, Baker FH (1975) The representation of the visual field in the lateral geniculate nucleus of Macaca mulatta. J Comp Neurol 161:569–594

Marko H (1969) Die Systemtheorie der homogenen Schichten. I. Mathematische Grundlagen. Kybernetik 5:221– 240

Nothdurft HC (1983) The influence of stimulus velocity on orientation tuning of striate cortex cells in cat. Neurosci Lett Suppl 14:S263

Payne BR, Berman N (1983) Functional organization of neurons in cat striate cortex: Variations in preferred orientation and orientation selectivity with receptive-field type, ocular dominance, and location in visual-field map. J Neurophysiol 49:1051–1072

Sawchuk AA (1974) Space-variant image restoration by coordinate transformations. J Opt Soc Am 64:138–144

Seelen W von (1968) Informationsverarbeitung in homogenen Netzen von Neuronenmodellen. Kybernetik 5:181–194

Türke B (1981) Analysis of pattern recognition by man using detection experiments. J Math Biol 13:47–65

Tusa RJ, Rosenquist AC, Palmer LA (1979) Retinotopic organization of areas 18 and 19 in the cat. J Comp Neurol 185:657–678

Wilson HR, Cowan JD (1973) A mathematical theory of functional dynamics of cortical and thalamic nervous tissue. Kybernetik 13:55–80

Two Views of the Cerebral Cortex

V. BRAITENBERG[1]

1 Introduction

The cerebral cortex, one half of the cerebral grey substance in mice and men, is what any detailed theory of the workings of the nervous system ought to explain, or at least, ought to make use of. In fact, theoretical papers ranging from 1943 to 1985 and from rather realistic views to frankly speculative constructs have made explicit reference to the cortex and perhaps even have influenced the ideas of some experimenters. Cortical anatomists and physiologists, in turn, learned to shape their findings so as to make them acceptable to the theoreticians. The resulting situation of reciprocal positive feedback had some stable solutions:

1. The random network with or without learning. Lashley's philosophy is of this category, as is Hebb's theory of cell assemblies. Rosenblatt's perceptron is also a descendant.
2. The circuit diagram in the spirit of radio engineering. The amplifier entered neurophysiology from communication engineering and with it came various ideas, the most enticing being that of functional secrets embodied in loops of wires connecting tubes, condensers and the like. The neuroanatomists responded quickly with loops of fibres connecting various sorts of neurons in the cortex (Lorente de No and others).
3. The digital computer and a logical theory of nerve nets. This was soon recognized as a misleading analogy, but the digital computer has at any rate among all models of cortical function the unique distinction of being a very useful machine. And the theory formulated by McCulloch and Pitts (1956), made more palatable by Kleene (1956), lent the brain a flair of almightiness which was gratefully recognized by many.

   I could mention some more, but I won't. Rather, it is important to realize that the greater part of brain research today ignores the mental schemata 1 to 3 and operates on the basis of more archaic pictures:

[1] Max-Planck-Institut für Biologische Kybernetik, Spemannstraße 38, 7400 Tübingen, FRG

Brain Theory
Edited by G. Palm and A. Aertsen

4. The idea of projection, derived from optics. Indeed, this is an enterprise that has not yet been exhausted: Wherever one looks, one finds maps of the body surface or of some sensory space repeated many times in the cortex or in subcortical structures. The brain is reluctant to give up spatial order when it corresponds to something meaningful in the outside world, even after many stages of elaboration. Whether this just reflects convenience of engineering, or some profound reason, we do not know. Projectionist research is still flourishing (see volumes by Woolsey 1981–1982).
5. The idea of localization of function, derived from dualist philosophy. If you are utterly convinced that the connection between mind and brain escapes us as matter of principle, all you can do is find out what part of the brain is related to which faculty of the mind, without asking any further questions about the nature of the connection. This leads to a very useful partition of the brain and to a perhaps less useful dissection of the psyche on anatomical grounds.

I will report in this paper on two different approaches to the cerebral cortex which we have been following in the past years, one in our own histological work on the mouse cortex (Braitenberg 1978a,b), the other one in an analysis of the papers by Hubel and Wiesel and their followers (Braitenberg and Braitenberg 1979, Braitenberg 1983, 1984). Our own work results in a view of type 1 above, while the Harvard papers reflect an extreme type 5 position, possibly with the implied hope of finding a type 2 explanation. Some of the most stunning findings in Hubel and Wiesel's papers carry the principle of localization of function much further than anybody would have believed: "orientation specificity" varying systematically on the surface of the cortex in a succession of strips 30 μm wide, colour specificity being confined to patches not much wider than that (Hubel and Wiesel 1977, Hubel and Livingstone 1982).

In view of these findings it would seem difficult to sustain the essentially random-network philosophy which had grown dear to our hearts on the basis of many anatomical facts. However, I will show that it is possible to make the two points of view quite compatible with each other.

2 The Neuropil in the Mouse Cortex

All the essential features of the cerebral cortex which impress us in human neuroanatomy can be found in the mouse too, except of course for a difference in size by a factor 1000. It is a task requiring some experience to tell a histological section of the mouse cortex from a human one, if the Golgi staining technique is used, and no clues about magnification are provided. With electronmicrographs the task would actually be almost impossible. The task is easiest with a low power photograph of a Nissl preparation (cell body stain). In fact, at the most microscopical level, the components of the nerve tissue are quite similar in different animals and even in different regions of the brain. On the other hand, a more macroscopical view may readily reveal differences between one piece of brain and the other, the difference being essentially related to the statistics of the distribution of the various components in the tissue. The essential similarity of the neurons as they appear in Golgi preparations of the cortex of mouse and man, the similar shape of their dendritic and axonal ramifications, shows that a common principle is at work in the cerebral cortex of the two species. This principle is quite different from that governing the wiring in other parts of the brain: some of the neuronal shapes in the cerebral cortex are quite characteristic for that level of the nervous system and occur only there. We should like to know why and are looking for an interpretation which provides a good reason for the structural peculiarities of the cerebral cortex.

Before we approach the problem of the shape of the neurons in the cortex, we take a more global view and collect some quantitative data about the densities of the various components in the cerebral cortex of the mouse. The numbers, many of them only approximate, are assembled in Tables 1 and 2. Table 1 contains the raw data. The macroscopical measurements (a, b, c) were taken on frozen sections of formalin fixed material. The areas (a, b) were measured by relatively rough graphic methods. The density of synapses (d) is a quantity affected by a variety of experimental difficulties, and varies somewhat in different areas and layers. Our figure is slightly in excess of other published

Table 1. Measured quantities

a:	Surface area: $2 \cdot 120\,\mathrm{mm}^2$ (including hippoc.)
b:	Surface area: $2 \cdot 65\,\mathrm{mm}^2$ (eulaminate isocortex)
c:	Thickness: 0,8 mm
d:	Density of synapses (isocortex): $10^9\,\mathrm{mm}^{-3}$
e:	Percent type I-synapses: 85% (Wolff 1976)
f:	Density of axons (electronmicroscopy, layer I): $4\,\mathrm{km/mm}^3$
g:	Density of neurons: $2 \cdot 10^5\,\mathrm{mm}^{-3}$
h:	Distribution of cell-types (very rough estimate):
	70% Pyramidal, 10% Martinotti, 20% others
	Dendritic length per neuron:
i:	Pyramidal: 3–5 mm
j:	Martinotti: 2–3 mm
k:	Stellate: 4–6 mm
	Axonal length per neuron (not including fiber in white matter):
l:	Pyramidal: 3–6 mm
m:	Martinotti: 3–4 mm
n:	Stellate (large): 10–17 mm
o:	Afferent: 5 mm
	Relative axonal field density:
p:	Pyramidal: 10^{-5}
q:	Martinotti: 10^{-4}
r:	Stellate: 10^{-3}
t:	"Cross section" of basal dendritic tree, Py-neuron (see Braitenberg 1978b)
u:	Number of spines per dendritic length (Py-neuron): $1.5\,\mu m^{-1}$
v:	Synapses per spine: 1

Table 2. Deduced relations

α (b,c):	Volume isocortex: $2 \cdot 52\,\mathrm{mm}^3$
β (α,d):	Number of synapses: 10^{11}
γ (α,g):	Number of neurons (isoc.): $2 \cdot 10^7$
δ (β,γ):	Synapses/neuron: 5000
c (d,f,ς):	Synapses per length of axon: 1 every 1 to 4 μm
ς (h,l,m,n):	Density of axons: $1 - 4\,\mathrm{km\,mm}^{-3}$
η (i,u):	Number of spines/Py-neuron 5000
θ (g,h,η):	Density of spines $7 \cdot 10^8\,\mathrm{mm}^{-3}$
λ (p,s,t,u,ε):	Probability of 0,1,2...contacts from Py to Py (w_0, w_1, w_2...) at distance 100 μm: $w_0 = 0,9$ $w_1 = 0,09$ $w_2 = 0,004$ same at distance 10 μm: $w_0 = 0,1$ $w_1 = 0,27$ $w_2 = 0,27$ $w_3 = 0,18$

measurements (Cragg 1967) but corresponds to an approximation quoted by many. The distinction of Type I and Type II synapses becomes rapidly very convincing to anybody who had the opportunity to familarize himself with cortical electronmicrographs, but is quite difficult to render objective. In any case, the percentage quoted (Wolff 1975) is very close to our own appraisals in various samples of the mouse cortex. If, by the method of forced choice one classifies synapses according to the well-known criteria, one always ends up with a 4/5 majority, or even slightly higher, of the kind of synapses (Type I) which we like to assume, but never were able to prove to be excitatory. This is a remarkable fact which will detain us later.

The density of neurons (g) in the mouse is almost ten times that in the human cortex. This reflects a difference in the size of the neurons, not so much of their cell bodies but rather of their dendrites and axons which are longer and more ramified and hence occupy proportionally more space when the total number of neurons and their average distances are larger.

A quantity which reflects the complexity of the interactions in the tissue is axonal density (f); the total length of axonal segments in a unit volume. This can be obtained by stereological reasoning on the cross-sections of axons recognizable on electronmicrographs, or from Golgi preparations in which the axonal tree of various neuron types can be measured (l, m, n, o). The total axonal density can be obtained from these values if the neuronal density (g) and the differential distribution of different neuron types (h) are known (see Table 2, ς). Both methods yield approximately the same result.

Similar measurements can be made on dendrites. It can be seen from the table that the length of all the dendrites of one neuron (i, j, k) is of the same order as that of the axonal tree (l, m, n) and the dendritic density in the tissue is therefore only slightly less than the axonal density.

The remaining measurements are particularly useful if one wants to estimate the influence which the various neurons have upon each other. The relative axonal density (p, q, r) is the proportion of the axonal population contributed by one particular neuron within the territory of its axonal spread. It is interesting to note that even the densest axonal trees of the stellate cells (r) and of the specific thalamo-cortical afferents (not given in the table) represent only one in a thousand axons present within the confines of their termination. Equally interesting, the loosest axonal trees, those of the pyramidal cells (p) seem to be specially made for a wide distribution of signals from each cell.

The "cross-section" of a dendritic tree (t) is defined as the probability of hitting one of the dendrites for a straight fibre entering the region of the dendritic tree in a random direction. For pyramidal cells, which have most of their afferent synapses on dendritic spines, this cross-section was measured on tracings of the dendritic tree including the spines. A closed envelope of the projection connecting all the spine tips was drawn. The area of this envelope, divided by the area of the entire dendritic expansion (i.e. the macroscopical envelope connecting the tips of the dendrites, or alternatively, the circumscribed circle) provides a measure for the probability termed "cross-section".

Finally, the last two quantities were important for assessing the importance of dendritic spines: the density of spines per unit length of (pyramidal neuron) dendrite (u) and the number of synapses on each spine (v), namely 1.

This crude list of facts, combined in various ways, provides some further quantities more directly relevant for theories of cortical function (Table 2). Besides the more trivial figures volume of the isocortex (α), total number of synapses (β) and of neurons (γ), we get the more interesting ratios. Since cortical synapses as a rule have only one presynaptic and one postsynaptic element we can easily compute the average number of synapses on the dendritic tree of cortical neurons (ς). The number of cortical synapses for which a cortical neuron is presynaptic is only slightly less, because the synapses provided by extracortical afferents are only a small fraction of all synapses, the vast majority being synapses between cortical neurons.

A very useful quantity is (ε), the density of synapses along an axon, which follows simply from the average number of synapses belonging to one neuron and from the average length of the axons of one neuron. There must be a synapse every 1 to 4 μm of axonal length, most of them "en passant", suggesting connections very different from the old picture of terminal boutons situated at the very tip of axonal branches.

The number of spines on all the dendrites of a pyramidal cell (η) is also a derived quantity, since spine counts are always performed on isolated segments of a dendritic tree. From this the density of the spines in the tissue can be derived (θ) which is only slightly less than the density of all synapses in the tissue.

In principle these figures provide the basis for a connectivity matrix of the cortical neurons, but of course the form factors reflecting the shape of dendritic and axonal ramifications are also involved. The probability of a connection between any two neurons in a block of tissue depends crucially on their distance. There are two ways of calculating this for two neighbouring pyramidal cells. One is based on the assumption that the connection is via straight axon collaterals (Fig. 1) and that the form factor on the receiving side is described by the

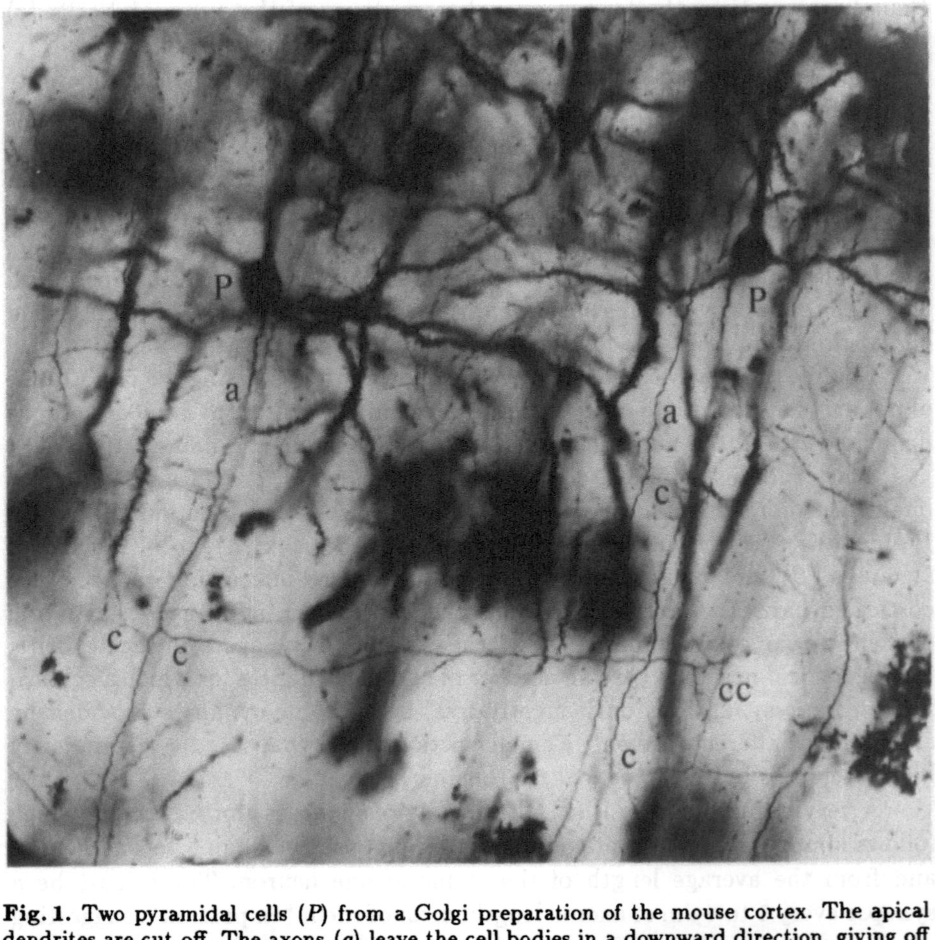

Fig. 1. Two pyramidal cells (*P*) from a Golgi preparation of the mouse cortex. The apical dendrites are cut off. The axons (*a*) leave the cell bodies in a downward direction, giving off straight collaterals of the first (*c*) and second (*cc*) order. Such collaterals are responsible for most of the synapses in the cortex. The target neurons are again other pyramidal cells. Due to their straight course, it is unlikely that such a collateral makes more than one contact with any particular neuron

"cross section" of the dendritic tree (t, Table 1). The other one takes as the axonal and dendritic form factor a certain density of pre- and postsynaptic sites homogeneously distributed within the territory of the axonal expansion of one neuron and the dendritic expansion of the other. If the overlap is known, and if we assume that the presynaptic elements pick their postsynaptic partners at random, we can compute the probability of a connection simply by means of the binomial distribution. The values obtained, for two different distances, are listed in Table 2, λ.

What are the propositions which can be extracted from these anatomical facts for the purpose of a physiological discussion? Four of them are presented in Table 3, and I will discuss their implications.

A and B: It seems that the vast majority of all synaptic contacts within the cerebral cortex are between one class of neurons, the pyramidal cells, and that these are of the excitatory kind, or at least histologically of Type 1, for such are all the synapses residing on the tips of the dendritic spines. This implies a sort of computation which takes us far away from the radio engineering analogy, discussed earlier as the philosophy of Type 2. It is not the combination of a number of components of different kinds, which is at the basis of cortical function, but rather an enormous collection of fundamentally similar neurons connected to each other by a huge number of contacts. The fact that these contacts are all excitatory (the inhibitory ones belong to the stellate cells which have a very different connectivity) is also worth considering. Clearly, for any non-trivial calculation one would be considerably restricted if no inhibitory interactions were available to implement the logical function of negation or negative quantities in arithmetic. But in an associative memory (Palm 1982) a multitude of excitatory contacts is what one would expect. This is our main reason for assigning this role to the cerebral cortex (Braitenberg 1978b).

Proposition C (Table 3) is also interpretable in the same vein. It states that the divergence of signals from one pyramidal cell is as large as it can be: the number of synaptic partners is almost as great as the number of synapses, each of the partners receiving just one or occasionally two synaptic contacts. Again, this is desirable in a network which ought to be prepared to discover, and store as "associations", the largest possible variety of correlations between the activities of the individual elements.

Table 3. Main propositions inferred from Table 1 and 2

A (d, h, θ):
Most synapses are on spines

B (h, A):
Most contacts are Py−Py

C (λ):
The number of neurons afferent to a given Py-Cell is almost as great ($\approx$ 5000) as the number of its afferent synapses. Similarly for efferent synapses

D (B,C):
The cortex is a mixing machine. The activity is relayed in ever new combinations from one set of neurons to the next

The upshot is proposition D: the cortex as a system of fairly uniform units, the pyramidal cells (with a smaller number of neurons of a different kind interspersed) connected to each other by wide-spread but very weak links, so weak, in fact, that the activity of one single cell can hardly exert an appreciable influence on the others. We must assume that the elementary event is constituted by fairly large sets of active pyramidal cells so that fresh sets will be activated in sucession through the synapses preformed in the cortical tissue and moulded by a learning process.

The picture I have in mind is influenced by Hebb's theory of Cell assemblies, by Abeles' "synfire chains", and by various developments of this work by Palm (1982).

There are two observations on cortical structure which are needed to complete the picture.

First, the shape of pyramidal cells. Their dendritic tree has an apical and a basal part, and also their axons are bipartite: long axons reaching distant pyramidal cells and axon collaterals making local connections. The local connections generally attach themselves to basal dendrites of other pyramidal cells, while the distant connections terminate on apical dendrites. We do not know whether the *A*-connections (apical dendrites, long axons) and the *B*-connections (basal-dendrites, axons collaterals) are two separate systems with different tasks, or just (apparently) different for reasons of convenience in the construction of the network during ontogeny. It is difficult to provide a very rich system of connections for millions or billions of neurons arranged in a cup-shaped volume, and one might well come up with the idea of keeping the shorter ones in the volume while letting the longer ones take shortcuts through the surrounding space. There may however be a more interesting distinction between the *A* and *B* systems according to Palm (Palm and Braitenberg 1979). While the probabilities of connections in the *B*-system clearly depend on distances in a smooth fashion, (the "metric system"), no such rule is apparent for the *A* system ("ametric system"). Thus one system may embody, or learn, facts of the world which refer to some metric spaces, (e.g. visual space) and the other facts referring to more abstract realms.

The second observation is about non-pyramidal cells. These are the (smooth) stellate cells, basket cells, chandelier cells in anatomical terminology. The general consensus is that they are inhibitory. The high density of their axons (Table 1, r) and especially their specialized endings (baskets, chandelier endings) imply that they have a much stronger grip on their target neurons than the excitatory pyramidal cells. Their role may be merely that of safety devices, strewn among the pyramidal cells to smother the local explosions which are to be expected in a system of overwhelming excitatory connections. But we may also assign them a more interesting role. What an associative memory learns is association of events, in other words positive correlations between the occurrence of events. There have been a number of suggestions of how this may be achieved by a "plastic" mechanism affecting the synapses. If the mechanism is supposed to discover, and translate into synaptic strength, negative as well as positive correlations, say, if the signal "A and B but not C" has to be learned as distinct from "A and B", then we are either driven to assume a much more complicated elementary learning mechanism than the additive one which is generally postulated, or we must use inhibitory interneurons for the negation. I suggest that at least some of the non-pyramidal cells, in particular those residing in the input layer IV have this function. The next part of this paper, where a special example will be discussed, makes use of this idea.

3 The "Wiring" of Area 17, the Primary Visual Area of the Monkey

Hubel and Wiesel (1977) discovered some facts by microelectrode recording which could be interpreted as indicating very specific connections within the cortical neuropil. The following is a selection of some of their most striking findings:

1. Most neurons in $A\,17$ respond best to patterns of light and dark in the visual field ("receptive fields") which are much larger, by a factor 10 or 20, than the distance between points "seen" by neighbouring retinal receptors.
2. The patterns which are most effective are those containing straight borders between light and dark regions presented at a certain fixed orientation characteristic of each neuron.
3. Proceeding through the cortex in any direction parallel to the cortical plane, one encounters neurons whose characteristic orientations vary continuously, with clockwise rotation of the orientations in the visual field sometimes changing smoothly into counterclockwise orientation and vice versa. There are also occasional abrupt changes of the preferred orientation for small displacements of the recoding electrode.

These three facts taken together constitute a puzzle when one tries to reconstruct the intracortical connections which would produce these strange exceptions to the apparent homogeneity of the cortical neuropil.

Figure 2 represents such an attempt. Two neurons, N_1 and N_2, situated in Area 17 at a distance of about $50\,\mu m$ from each other, have receptive fields F_1 and F_2 in the visual field (a). The fields are about 15 times as large as the distance between two neighbouring points of the elementary grid (dots) corresponding to the resolution of the visual field. Since the plane of the cortex represents the visual field, the two neurons N_1 and N_2 correspond to slightly different positions in the visual field, say to positions shifted by the angle of resolution, i.e. by the distance between neighbours in the elementary grid. Thus we draw F_1 and F_2 slightly apart (in reality there is considerable scatter in the representation of the visual field on most of the neurons in the cortex: the two

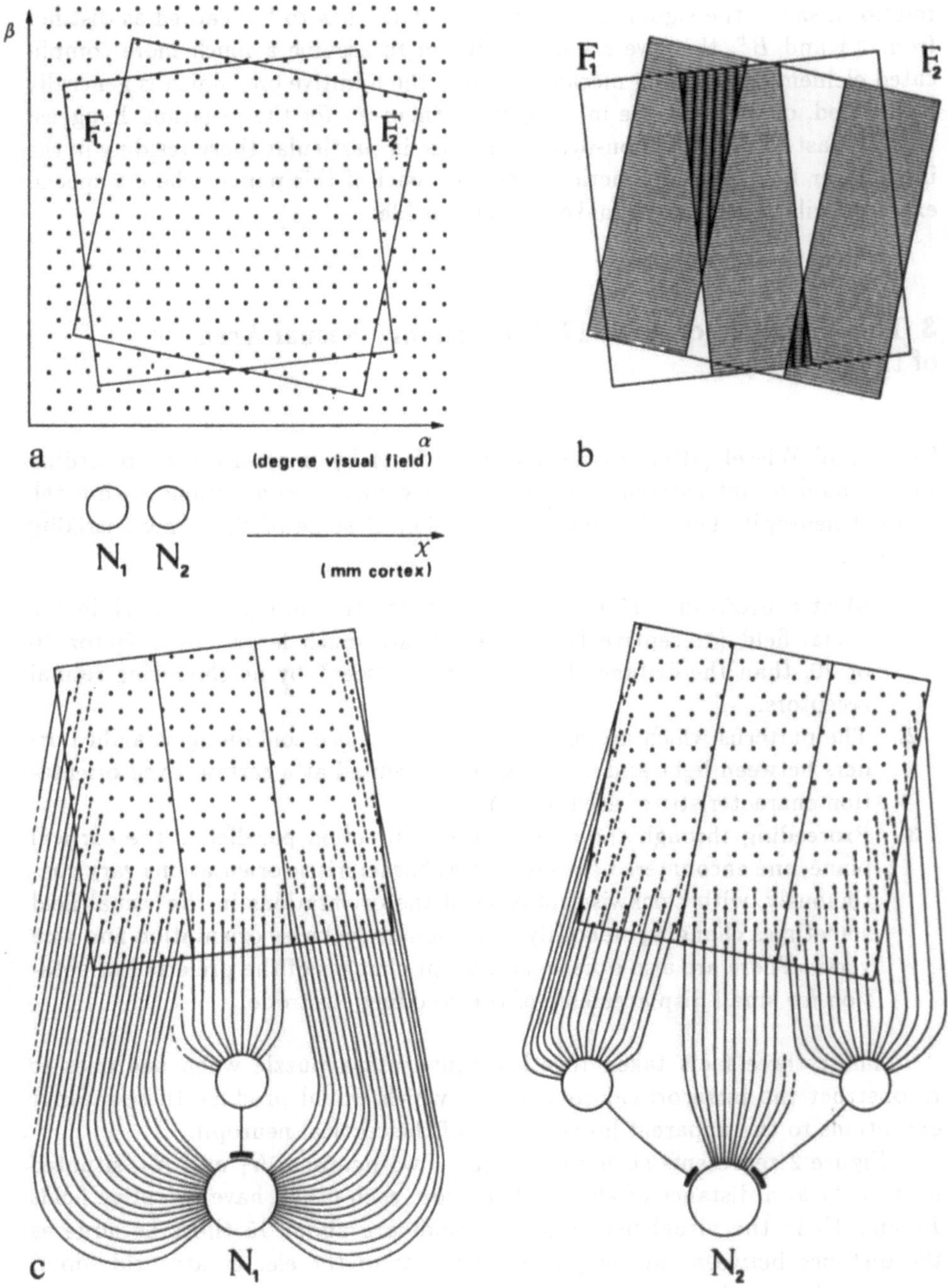

Fig. 2a–c. Illustration of a naive neuronal model explaining some of the effects described by Hubel and Wiesel. **a** The array of *dots* represents the sampling points in a portion of the visual field corresponding to individual cones of the retina. Their separation is about 1 min or arc. E_1 and E_2 are the two receptive fields of a pair of neurons N_1 and N_2 situated in the cortex at a distance of 50 μm from each other. The two fields are rotated 18° one with respect to the other (they belong to different "orientation columns"), and shifted by one minute of

fields F_1 and F_2 could be much farther apart, or entirely superimposed). On the other hand, since distance in the cortex also represents orientation in the visual field, we draw the receptive fields F_1 and F_2 rotated 18° one with respect to the other, for that is roughly the rotation corresponding to 50 μm cortex in many of the published records.

In Fig. 2b the two fields F_1 and F_2 are redrawn with their excitatory (white) and inhibitory (hatched) subfields. This is a feature which is found in many of the receptive fields: in different parts of the field a spot of light may produce an increase or a decrease of the activity of the corresponding neuron. F_1 has a central inhibitory region flanked by two excitatory regions, F_2 the other way round. Figure 2c shows the cortical "wiring" which one would naively assume to be an explanation of the response characteristic of the neurons N_1 and N_2. The two sets of fibres are drawn separately for the two neurons, since their superposition would make the diagram completely incomprehensible. We have to assume at least one inhibitory interneuron for N_1, two for N_2.

There are many more (by a factor of at least 10) neurons in Area 17 than there are incoming fibres from the geniculate body. It is clear that the complete blueprint for all cortical neurons in any small region of cortex according to the principle of Fig. 2c is not only an impossible task for the artist, but a highly improbable feat for the mechanisms of embryogenesis. If such improbable wiring were indeed preformed in the cortex (and that it is not acquired through learning we know from certain experiments: Hubel and Wiesel 1977), we would have to assume an amount of genetic information in the neuropil much richer than one would expect, having accepted the description of the mouse cortex in the first part of this paper.

But there is an alternative explanation of the Hubel and Wiesel effects requiring much less specificity of growth in the cortex (Braitenberg and Braitenberg 1979, Braitenberg 1984, 1985). The main idea was derived from a geometric analysis of the published records showing the variation of orientation specificity of cortical neurons encountered along a straight electrode track. It seemed that the records were more compatible with an inhomogeneity of the cortical neuropil circularly symmetric around centres spaced about half a millimetre apart, rather than with a local anisotropy, the orientation of which changes in one direction only ("orientation strips"), as Hubel and Wiesel had supposed. When this idea was first formulated (Braitenberg and Braitenberg 1979), nothing was known in the histology of Area 17 to justify the assumption of centres with a regular geometric arrangement, but soon afterwards the dis-

arc, the angle corresponding (in the fovea) to 50 μm of cortex. **b** The two fields F_1 and F_2 happen to be of the "simple cell" variety, with well-defined excitatory (*white*) and inhibitory (*cross-hatched*) subfields. E_1 has a central inhibitory region and two excitatory flanks, E_2 vice versa. **c** The wiring responsible for the receptive field characteristics of N_1 and N_2 is shown. Fibres from the receptors (*dots*) reach the cortical neurons directly in the excitatory subfields, via an interposed inhibitory neuron (*smaller circles*) in the inhibitory subfields. The two diagrams of Fig. 2c should be mentally superimposed, to explain the effects described in a and b covery of the "cytochrome oxidase blobs" (Horton and Hubel 1981, Humphrey and Hendrickson 1980) regularly arranged with the spacing predicted by us came as a pleasant surprise.

However, the coincidence remains simply phenomenological until we are able to show in what way something concentrated in patches in the visual cortex can produce the effects described by Hubel and Wiesel. I offer the following hypothesis (Braitenberg 1983, 1984), best explained on the drawings of Figs. 3 and 4. I assume that the cytochromeoxidase blobs are the site of special inhibitory neurons which exert a strong inhibitory influence on all the (principally pyramidal) neurons in the surrounding region. I also assume that these inhibitors are connected, like the other neurons, to the input fibres in a strictly topographical fashion. The dendritic fields of the inhibitors which receive the input are circular with a diameter slightly less than their separation. The inhibition is so strong that a uniform excitation of a large area is completely smothered by the inhibitors, Fig. 4a. Clearly then only elongated stimuli can be effective which fit into the interstices between the inhibitors so as not to activate them. Moreover, such elongated stimuli have to be oriented tangentially with respect to a neighbouring inhibitor, in order not to affect its dendritic field (Fig. 3). How long straight patterns of excitation affect an array of "hypercolumns", each dominated by a central inhibitor, depends on the orientation of the stimulus with respect to the axes of the geometrical array (Fig. 4). The orientation of the effective stimulus tangential to circles surrounding the hypercolumn centres produces the well-known electrophysiological effects of smooth sequences of orientation along a microelectrode track, including the switching

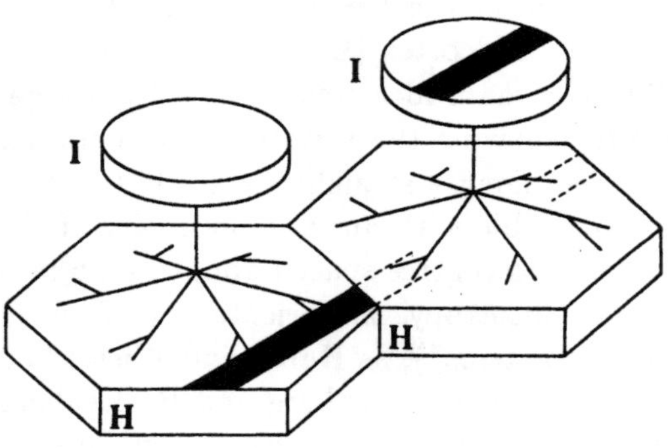

Fig. 3. An alternative model of orientation specificity. The cortex is composed of (schematically) hexagonal "hypercolumns" *H*. In the centre of each hypercolumn there is an inhibitory region, here represented above the hypercolumn as a round box *I*, perhaps to be identified with the dendritic field of an inhibitory neuron. When the central inhibitory region is hit by the stimulus, here represented as a black bar, the corresponding hypercolumn is inhibited and does not respond to the stimulus (*right* hypercolumn). The stimulus is effective only if it passes by the inhibitory region (*left* hypercolumn). Clearly, the hypercolumns respond more readily to elongated stimuli orientated tangentially with respect to the hypercolumn centre, for the same stimuli with a radial orientation tend to hit the inhibitory hypercolumn centre. In this fashion different regions of the periphery of each hypercolumn become sensitive to different orientations tangentially arranged with circular symmetry around the centre

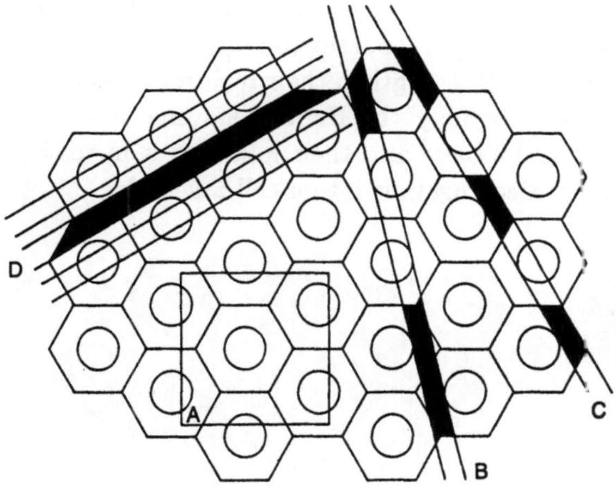

Fig. 4. An array of hypercolumns as in Fig. 3 seen from the top. Regions where pyramidal cells respond to the stimuli *A* to *D* are shown in *black*. The square stimulus *A* does not elicit any response since it falls on all the central inhibitors (*circles*) of the hypercolumns which it touches. *B* and *C* are bars of different orientation. They produce excitation in some hypercolumns where they bypass the inhibitory centres. The position of the active cells within the individual hypercolumns is different for *B* and *C* and is characteristic of the orientation (different regions of each hypercolumn have different orientation specificity). In *D* the stimulus is composed of three parallel stripes. The central one passes between two rows of hypercolumn centres and therefore elicits a continuous response in all the hypercolumns it touches. The two flanking stripes fall on rows of inhibitory regions. This arrangement is possible when the orientation of the stimulus is parallel to one of the axes of the array of hypercolumns. Neurons which respond to such stimulus configurations are called "simple cells" in the Hubel and Wiesel terminology. In our model they should have predetermined orientation in each part of the visual field. Other neurons responding to elongated stimuli with orientations such as in *B* and *C* would be "complex cells"

of the direction of orientation change from clockwise to counterclockwise, and the occasional abrupt change of orientation (Braitenberg 1985).

The model of Figs. 3 and 4 is simplified in many ways but correct in principle. One simplification regards the geometrical pattern of the blobs, which is not so regular in reality. Also, the model fails to distinguish between two kinds of input fibres, "on"-centre and "off"-centre, each already with its structured receptive field, one the negative of the other, one activated by light in the centre and depressed in the periphery, the other one vice versa. In spite of this and other shortcomings, the drawings make the main point, namely that the anisotropy determined by the hypercolumn centres, if they are the site of special neurons, is sufficient to explain the known effects without having to resort to a great deal of highly specified wiring. In fact we did not assume anything but circularly symmetric, perhaps largely random dendritic and axonal distributions of the inhibitory neurons.

We have not yet mentioned the wiring of the other cells, which we assume to be pyramidal cells. They are strongly connected to each other by excitatory connections subject to "plastic" changes in the way of associative learning. Clearly, neurons of the same or of neighbouring hypercolumns which are tuned to the same orientation will often be activated by the same stimuli in the visual field either at the same time or in close temporal succession. Such neurons, following the laws of associative learning, will strengthen their reciprocal synapses and will form what was called a "cell-assembly" by Hebb. I assume that a receptive field as described by Hubel and Wiesel and their followers is in reality the compound field of many neurons tied together into a cell assembly.

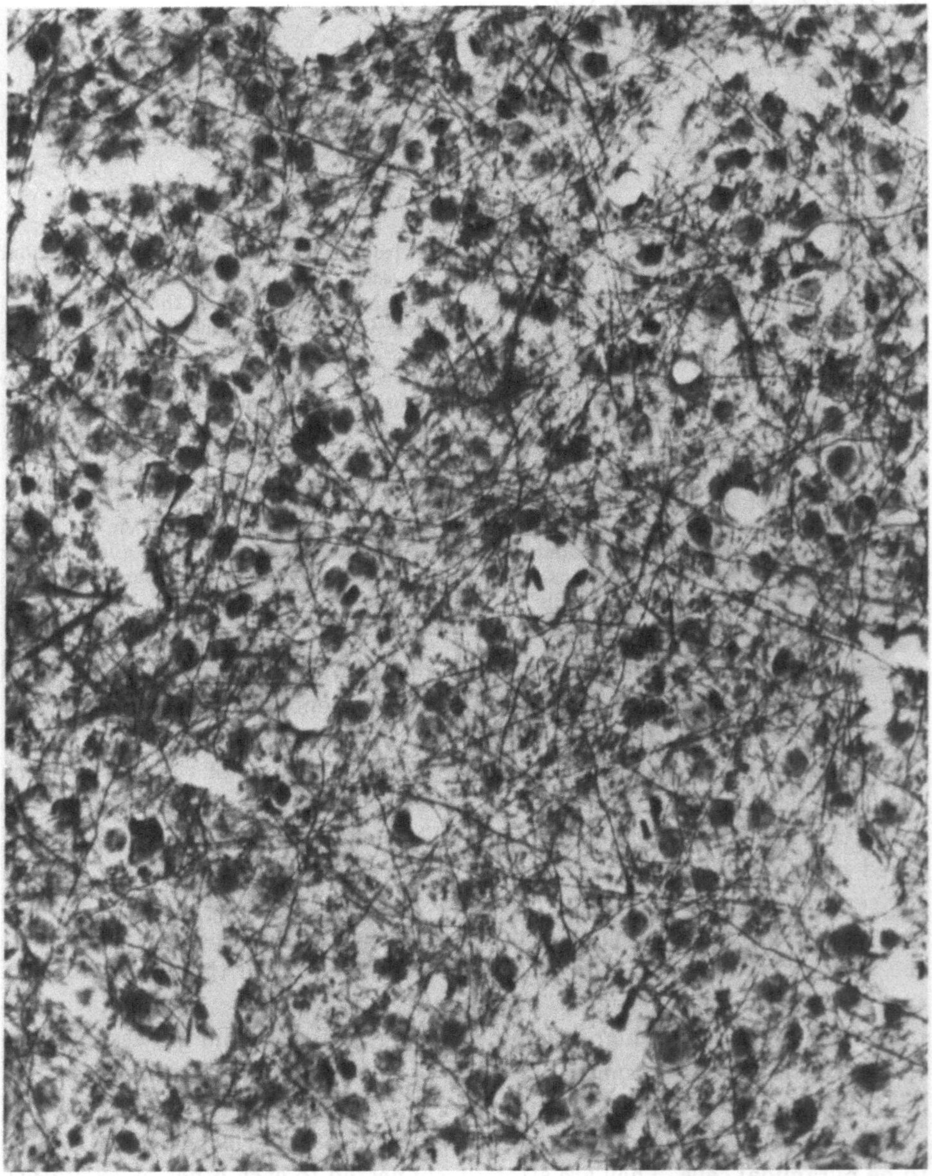

Fig. 5. Horizontal section through layer IVb of Area 17 of the monkey. The orientation of fibres in the anatomical picture appears random, contrary to naive expectation

This is a necessary assumption in our model, which would assign to individual neurons only small fields no larger than a hypercolumn (translating the cortical coordinates into those of the visual field). In reality, the typical field is two or three times hypercolumn size.

The model makes a strong prediction. In Fig. 4 it is obvious that what is called a simple cell in Hubel and Wiesel terminology, namely one with definite parallel excitatory and inhibitory subfields should be oriented along one of the axes of the array of hypercolumns. Cells with receptive fields not so oriented would be "complex". If in a small area of cortex simple and complex cells are sampled, the orientation of the respective receptive fields should be complementary.

Again, the wiring of the pyramidal cells subserving the associative learning and hence the formation of the receptive fields in this model of monkey area 17 is no different from what we had derived from our statistical analysis of the neuropil in the mouse cortex. Thus I have fulfilled my promise of tying together two descriptions of the cortex that initially seemed almost incompatible. A tangential section through layer IVb of the monkey striate cortex (Fig. 5) showing fibres running in all directions is apt to underscore this point.

References

Braitenberg V (1978a) Cortical architectonics: General and areal. In: Brazier MAB (ed) Architectonics of the cerebral cortex. Raven Press, New York, pp 443–465

Braitenberg V (1978b) Cell assemblies in the cerebral cortex. In: Heim R, Palm (eds) Lecture notes in biomathematics, vol 21. Springer, Berlin Heidelberg New York, pp 171–188

Braitenberg V (1983) Explanation of orientation columns in terms of a homogeneous network of neurons in the visual cortex. Soc Neurosci Abstr vol 9

Braitenberg V (1984) Das Raster der neuralen Elemente in der Sehrinde und seine neurophysiologischen Konsequenzen. In: Herzau V (ed) Pathophysiologie des Sehens. Enke, Stuttgart

Braitenberg V (1985) Charting the visual cortex. In: Peters A, Jones EG (eds) Cortex, vol III pp 379–414

Braitenberg V, Braitenberg C (1979) Geometry of orientation columns in the visual cortex. Biol Cybern 33:179–186

Cragg BG (1967) The density of synapses and neurones in the motor and visual areas of the cerebral cortex. J Anat 101:639–654

Horton JC, Hubel DH (1981) Regular patchy distribution of cytochrome oxidase staining in primary visual cortex of macaque monkey. Nature (London) 292

Hubel DH, Livingstone MS (1982) Cytochrome oxidase blobs in monkey area 17: Response Properties and afferent connections. Neurosci Abstr 8:706

Hubel DH, Wiesel TN (1977) Ferrier Lecture. Functional architecture of macque monkey visual cortex. Proc R Soc London Ser B 198:1–59

Humphrey NK, Hendrickson AE (1980) Radial zones of high metabolic activity in squirrel monkey striate cortex. Soc Neurosci Abstr 6:315

Kleene SC (1956) Representation of Events in Nerve Nets and Finite Automata. In: Shannon CE, McCarthy T (eds) Annals of mathematics studies no 34. Princeton Univ Press, London

McCulloch WS, Pitts WH (1943) A logical calculus of the ideas immaent in Nervous Activity. Bull Math Biophys 9:127–247

Palm G (1982) Neural assemblies. Studies of brain function, vol VII. Springer, Berlin Heidelberg New York

Palm G, Braitenberg V (1979) Tentative contribution of neuroanatomy to nerve net theories. In: Trappl R, Klir GJ, Ricciardi L (eds) Progress in cybernetics and systems research, vol III. Wiley, New York, pp 369–374
Wolff JR (1976) Stereological analysis of the hetrogeneous composition of central nervous tissue: synapses of the cerebral cortex. In: Underwood IA (ed) Proc 4th Int Congr Stereol, Sept 4–9, Gaithesburg. Natl Bur Standards, Washington, pp 331–335
Woolsey CN (1982) In: Woolsey CN (ed) Cortical sensory organization, vols I–III. Humana Press, Clifton

EEG Spatial Pattern Differences with Discriminated Odors Manifest Chaotic and Limit Cycle Attractors in Olfactory Bulb of Rabbits*

W.J. FREEMAN[1] and G. VIANA DI PRISCO[1,2]

1 Introduction

The immediate aim of this study was to find reproducible but differing patterns in the EEG of the olfactory bulb that recur with presentation of different odors. The underlying purpose was to use these patterns to explain how the bulb operates on its receptor input in the process of olfactory discrimination.

Rabbits that are trained to respond by licking to one odor with reinforcement (CS+) but not to another odor without reinforcement (CS–) tend to sniff each odor before making a discriminative response. By inference, between the onsets of the sniff and the lick response the olfactory bulb has already received the input from the receptors activated by the CS+ or CS– odor, and it is generating neural activity for transmission to the olfactory cortex that subserves discrimination between the two odors as the basis for response selection. Even if the odor discrimination is not completed in the bulb, the requisite information must be contained in differing bulbar neural activity patterns. If the EEG adequately manifests those activity patterns, then a distinctive and reproducible difference should exist in some aspect of the EEG during presentation of each discriminated odor (Viana Di Prisco and Freeman 1985).

In this study the EEG is decomposed by filters into various components, and each is tested for its efficacy to classify bursts correctly to preceding control, CS+ odor, and CS– odor periods. The procedure is analogous to a process of chemical purification, with the fractions tested for efficacy after each separation. Eventually a fraction is found, which serves optimally to classify EEG records correctly as belonging to CS+ or CS– trials. This report describes briefly the separation process, the test of efficacy, the distinguishing characteristics of the EEG, and the implications for the neural function of the bulb in sensory processing.

*Supported by a grant MHO6686 from the National Institute of Mental Health and by a grant NS16559 from the National Institute of Health

[1] Department of Physiology-Anatomy, University of California, Berkeley, CA 94720, USA

[2] Present address: Catedra de Fisiologia, Escuela de Medicina J.M. Vargas U.C.V., Esquina San Lorenzo, San Jose, Caracas 1010, Venezuela

Brain Theory
Edited by G. Palm and A. Aertsen

In regard to mechanism the bulbar neurons form an interactive neural mass that is distributed in the two dimensions of the bulbar surface. Interactions include mutual excitation, mutual inhibition and negative feedback between mitral (excitatory) and granule (inhibitory) cells (Freeman 1975). Owing to the lengths of basal dendrites and axon collaterals running parallel to the surface the interactions are diffuse and widespread. To a good approximation this distributed system can be described as a set of coupled oscillators (Freeman 1979). Owing to nonlinearities in the dynamics of the component neurons, we can predict that the bulb has multiple stable states. Three classes of these states are identified for this type of system (Garfinkel 1983): equilibrial, limit cycle and chaotic. The types of output characteristic of these three states are respectively tendencies to go to rest, to enter periodic oscillation or to sustain aperiodic, pseudorandom activity. The time-series output in an equilibrial state under perturbation from noise may closely resemble chaotic activity but must be distinguished by its degrees of freedom. The further aim here is to classify the EEG patterns observed before and during odor stimulation into these three types. Their occurrence or absence indicates the presence or absence of equilibrium points, limit cycles and chaotic domains, that constitute the attractors (Garfinkel 1983, Abraham and Shaw 1982) in the state space of the bulbar neural mechanism for odor identification.

This theory provides a necessary basis on which to analyze the EEG data and to explain how the neural mechanism of the bulb discriminates odors.

2 Methods

2.1 Establishing the Data Base

In a previous study (Viana Di Prisco and Freeman 1985) thirsted rabbits were trained to respond with licking to one odor (the CS+) paired with intraoral water delivery, and not to lick on presentation of another odor (the CS–). The rabbits tended to sniff to both odors but more frequently to the CS–. In a set of six sessions four of five rabbits acquired strongly the CR+ (licking with or without sniffing) and half as strongly the CR– (sniffing without licking) in the first two sessions and maintained these levels of responding thereafter. The fifth rabbit often sniffed but seldom licked and then indiscriminately to both the CS+ and CS–.

The EEG was recorded from the olfactory bulb by a 64-channel array of electrodes (8×8, 4×4 mm) surgically placed on the lateral surface. When the behavioral pattern was stabilized by training, the bulbar EEG underwent a reproducible change on each trial after delivery of an odor and after onset of the sniff response but before onset of the lick response (if it occurred) and before the UCS presentation (if any). In this change the EEG amplitude decreased by 35% on the average over subjects, the frequency on the average decreased, and the spatial patterns of root mean square (rms) amplitude changed slightly but significantly.

The bulbar EEG consisted of a recurrent burst oscillating at a frequency between 20–100 Hz for about 0.1 s with each inhalation and riding on the crest of a slow wave synchronous with respiration, usually at $3-5\,s^{-1}$. On each trial a 3 s control record was immediately followed by a 3 s test record during which the CS was delivered and the CR+ and CR– were sought. After each trial the EEG records were displayed on an oscilloscope; three bursts were selected from the control period (labelled C1, C2, and C3 starting from the end of the control period), and three bursts were taken from the test period (labelled T1 between odor and sniff onsets and T2, T3 between sniff and lick onsets). Each stored segment contained 64 EEG traces digitized at 2 ms intervals for a duration of 76 ms.

Each session contained 10 CS+ and 10 CS– trials randomly interspersed. The data summarized here were taken from sessions four to six when response patterns were stable. Each rabbit yielded 360 bursts: 180 control bursts and 90 CS bursts with each odor. Further details have already been published (Viana Di Prisco and Freeman 1985).

Each EEG was amplified (10 K), filtered (3db fall-off at 10 Hz and 300 Hz), digitized (8 bits/read, 10 ms read time, and 2 ms interval between time frames), and stored in a 64×38 matrix. The data on each channel were corrected for small differences in preamplifier gain, digitally smoothed by averaging over adjacent time samples, and detrended to remove the remaining respiratory wave. Bad channels (usually 2 to 4 and not more than 10) were replaced by the average of adjacent pairs. The correlation matrix of each trace with the burst ensemble average was computed; traces with correlation coefficients less than 0.20 were similarly replaced (an average of 2.2/burst), because by inspection of records these non-correlated waves were either from edge electrodes contaminated with nonbulbar EEG or were obscured by transient noise. The ensemble average $E(t)$ was then reformulated as the best available estimate of the EEG time course of each burst.

2.2 Decomposing the EEG Burst

Each burst was decomposed by the combined use of spectral analysis and curve-fitting. The most effective of several procedures was as follows (Freeman 1985a). The Fourier transform was taken of the ensemble average $E(t)$. The frequency and phase were identified for the highest gain; these values were used as initial guesses to fit a cosine wave to $E(t)$ in order to determine the amplitude V, frequency f, and phase p coefficients with a criterion of least squares deviation. The equation was

$$v(t) = \sum^{i} [1 + AM_i(t - t_m)] V_i \cos[FM_i(t - t_m)t + 2\pi f_i t + p_i],$$

where t_m was the midpoint of each trace, AM_i was the fixed amplitude modulation coefficient, FM_i was the fixed frequency modulation coefficient (both modulations linear with time) and $i = 1$ for the 1st component to $i = 5$ for the 5th. This first curve was subtracted from $E(t)$. The procedure was repeated on the interim residue 4 times, leaving a final residue of less than 3% of the energy of $E(t)$. Then the 5 frequency and 5 phase values were fixed, and each of the 64 waves was fitted by linear regression with the sum of the 5 cosine waves. This yielded 5 8×8 matrices of amplitude coefficients that represented the spatial patterns of the 5 components, and an 8×8 matrix of the rms residuals, estimated to contain an average of 19% of the total energy of control bursts and 25–27% of odor bursts, that had lower amplitude.

Each 8×8 amplitude matrix was subjected to a two-dimensional spatial Fourier transform. A low-pass 4th-order exponential filter (Gonzalez and Wintz 1977) was applied to the spatial frequency matrix with its 3 db fall-off set at $0.5\,c\,\text{mm}^{-1}$. This operation served to remove activity that was generated at spatial frequencies above the spectral range accessible to the bulbar granule cell generator (Freeman 1978, 1985b). The amount of energy removed from the total energy in the components was estimated to be on the average $16 \pm 2\%$ in control bursts and $23 \pm 3\%$ of odor bursts. The inverse Fourier transform was applied to take the filtered data back to the spatial domain.

The next operation was spatial deconvolution by the "software lens," which served to compensate for the distortion of the granule cell field potential by the volume conductor properties of the bulb (Freeman 1980). The depth of focus was set at 0.49 mm by systematic search to find the optimal value, a few microns less than the average depth of the granule cell generator in these rabbits estimated postmortem from the depth of the mitral cell layer. The procedure was applied to each of the 5 amplitude matrices in turn.

In an alternative procedure the 64 values of phase were determined from the Fourier transform of each wave form or by curve-fitting using nonlinear regression. The phase matrices were not useful for burst classification and are not further considered here.

The final steps were to correct the amplitude and phase values at each frequency for the attenuation and phase distortion caused by bandpass filtering, smoothing, and the digitizing time lag over the 64 channels, and to normalize the $360 \times 64 \times 5$ amplitude values for each subject across sessions by channel to zero grand mean and unit standard deviation for each channel prior to statistical analysis (Freeman 1985c).

3 Results

3.1 The Distributions of Spectral Energy

Some representative examples of the ensemble average are shown in Fig. 1 together with the Fourier transform, at the stage of identification of the first component and calculation of the interim residue, illustrating the multiple subsidiary peaks found in the spectrum. The mean values for the relative energy

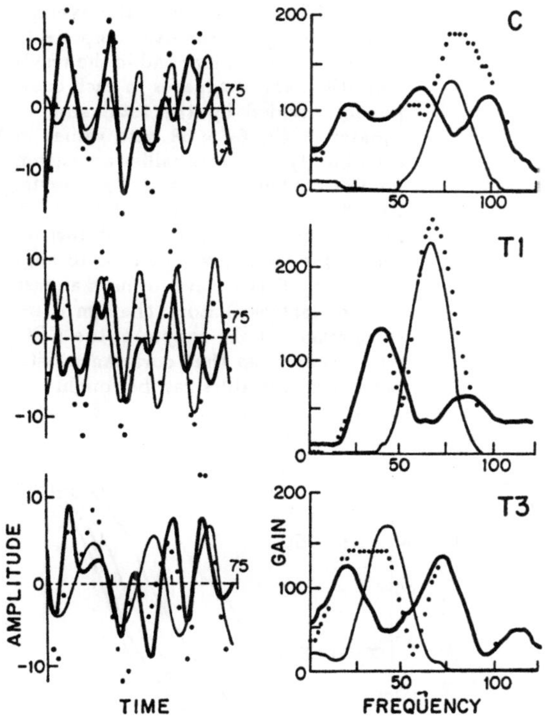

Fig. 1. Three examples are shown of the decomposition of bursts from one subject and trial. *Left* the *dots* show the digitized ensemble average; the light curves show the fitted dominant components; the dark curves show the residuals after the first subtraction. *Right* the Fourier transform gives the spectral gains of the data, the dominant component and the residuals. *C* control bursts; *T1* first odor burst; *T3* third odor burst. Time in ms, frequency in Hz. Freeman (1985a)

(the sum of 64 × 38 squares of each component divided by the total sum of 64 × 38 squares of the EEG) in each of the 5 components and the residuals for the 64 traces in each burst are shown in Fig. 2.

Figure 3 shows the variation over bursts in the energy distribution among components. The left-most curve is for the first component; the second is for the sum of the first two, and so on. The abscissa shows the relative energy content; the ordinate shows the proportion of 900 bursts having the energy content at each level. The compilation was done separately for control and odor bursts to show the tendency for relative energy to be more broadly distributed beyond the first component in the odor bursts.

The distribution of frequencies is shown in Fig. 4 for both control and odor bursts having dominant frequencies (f_1 in equation in Methods) > 55 Hz for all subjects (upper frame) and one exemplary subject (lower frame). The 5 values from each burst tended to spread across the frequency range from 10 to 125 Hz; the mean interval was 21.1 ± 5.3 Hz (standard deviation). Not every frequency subrange of 20 was represented in each burst; the interval was then greater by a factor of about 2 about the missing frequency value. The weighted mean frequency was 68.7 ± 7.6 Hz for the control bursts and 61.5 ± 5.3 Hz for the odor bursts, reflecting a small but unequivocal decrease in average burst frequency with odor presentation.

The spectral distributions of the energy are shown in Fig. 5. The dominant frequency of most bursts shifted randomly within the spectral zone of 55 to

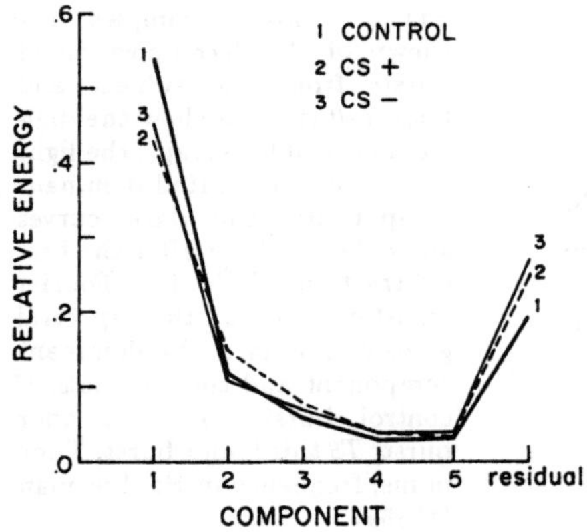

Fig. 2. The curves show the average relative energy in the five components ranked in order of size and in the residuals (the sum of squares in each component divided by the total sum of squares of the 64 × 38 data values in each burst). Separate tallies are shown for the control, *CS+* and *CS−* bursts. Decomposition by curve-fitting is an artifice for measurement and further processing; it is not intended to suggest that all the wave-forms manifest independent oscillators; the aim is incorporation of the shapes of the EEG into numbers so that covariance with behavioral variables can be sought

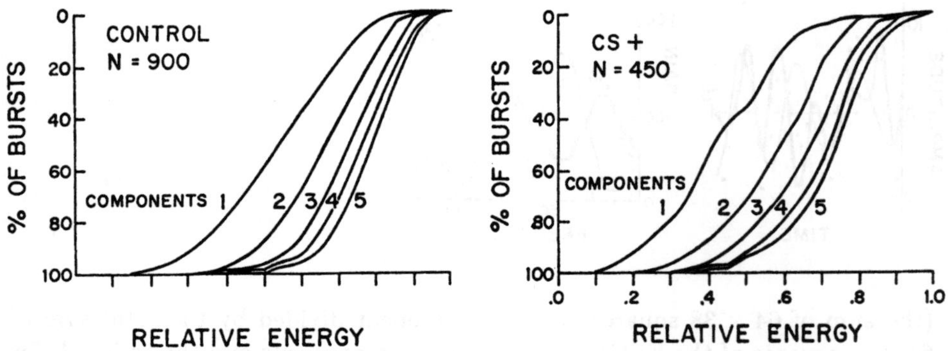

Fig. 3. Variation in energy distribution is shown for 900 control bursts and for 450 *CS+* odor bursts. The abscissa shows the energy of the dominant component (*1*) and the increased amount with the addition of othe components (*2–5*). The ordinate shows the percentage of all bursts having at least the designated amount of energy included in the specified number of components. *Left frame*: control bursts from five rabbits over three sessions; *right frame* odor bursts for *CS+*. (Freeman 1985a)

75 Hz. Comparing the control and odor bursts, upon odor arrival there was a substantial decrease in the energy in the 55–75 Hz region. On the average there was an increase in the energy in the 20–50 Hz region, both in the relative fraction of total energy (averaging 2 fold) and in the absolute level of energy (averaging +12%), partially off-setting the average 35% decline in burst rms amplitude upon odor presentation. Despite this spectral shift the slight majority of odor bursts (60%) had dominant components in the 55–75 Hz range. This average shift was localized to a subset of odor bursts, which had components in the frequency ranges > 55 Hz and < 55 Hz, but in which the energy in a lower frequency component was greater than in a higher frequency component (shaded histogram in Fig. 5), as opposed to the more common ratio (clear histogram). This type of burst was found in approximately 40% of odor bursts of both kinds, and also in about 10% of control bursts.

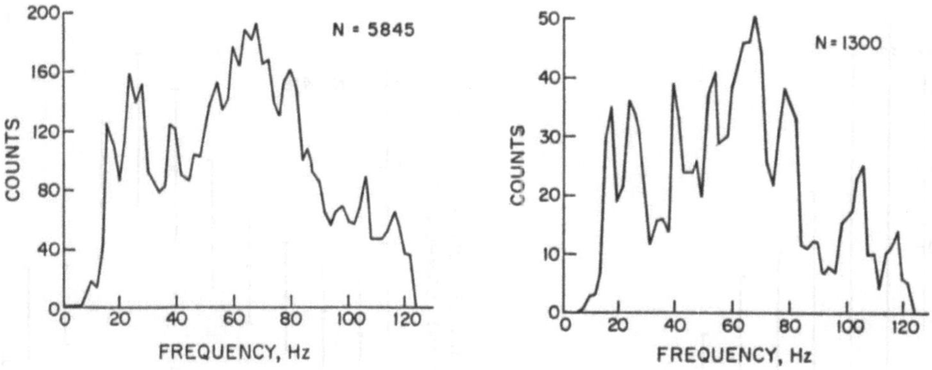

Fig. 4. *Left frame* shows an example of the distribution of the frequencies of the five components of 1169 bursts having $f_1 > 55$ hz from five subjects. *Right frame* is an example from one subject. The limits of the spectral distributions near 10 Hz and 120 Hz were determined by filtering. Within these limits there was no evidence from any rabbit for fixed frequencies or integer multiples to suggest harmonics. The mode of the intervals between components was 19 Hz

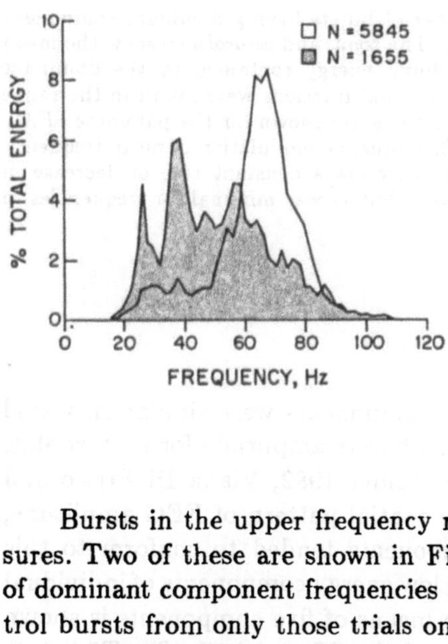

Fig. 5. Weighting each frequency count by the relative energy of the component gave an energy distribution that approximated the average power spectrum without the residuals. The values are shown separately for bursts with $f_1 > 55$ (*clear*) and $f_1 < 55$ Hz (*shaded*). The latter type of burst occurred four times more often in test periods than in odor periods. (Freeman 1985a)

Bursts in the upper frequency range were more coherent by several measures. Two of these are shown in Fig. 6. The upper frame shows a histogram of dominant component frequencies (denoted f_1) over the set of odor and control bursts from only those trials on which a correct CR occurred. The mean and standard deviation of the fraction of power in the dominant component of each burst are shown with its frequency f_1 in the ranges indicated. The maximal concentration of energy was in the upper frequency range. The lower frame shows the means and standard deviations of the frequency modulation parameter FM_1 (see Methods) for f_1 similarly grouped. There was a tendency for frequency slowing ($FM_1 < 0$) in bursts with $f_1 > 60$ Hz, and for acceleration ($FM_1 > 0$) in bursts with $f_1 < 60$ Hz. The variance in FM_1 increased exponentially with decreasing f_1 frequency.

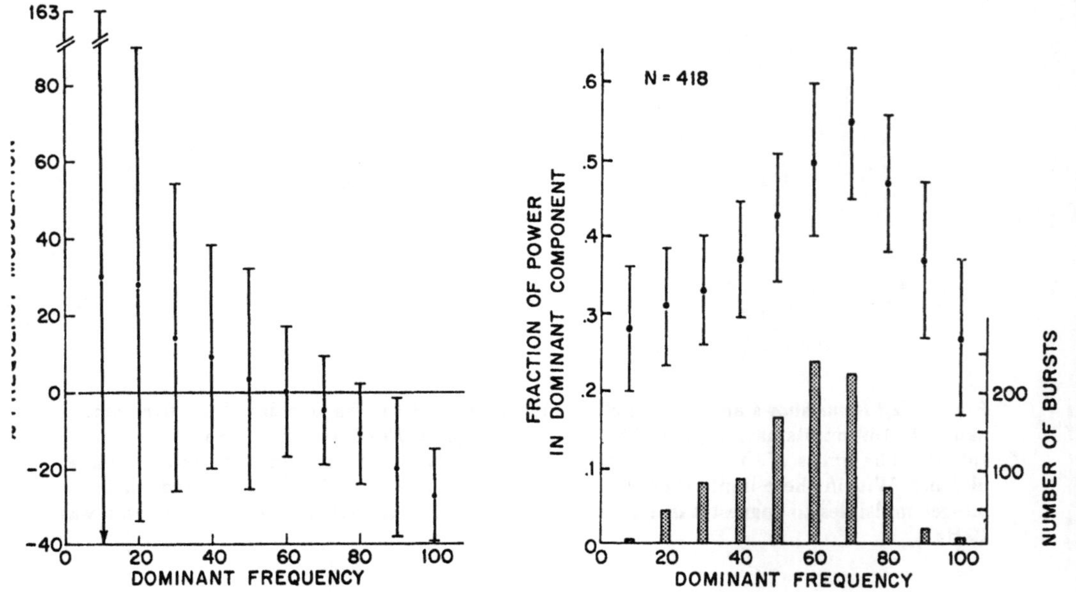

Fig. 6. *Right frame:* the *shaded bars* show the number of bursts having dominant component frequencies f_1 in the indicated frequency intervals. The *points* and *vertical lines* show the mean and standard deviation of the fraction of total burst energy contained by the dominant component in each designated frequency range. Maximal fractions were found in the range 55–75 Hz. *Left frame* the means and standard deviations are shown for the parameter FM_1 (equation in Methods). This is expressed here as % frequency modulation of mean frequency f_1 of the dominant component. Negative values represent a constant rate of decrease in frequency with time. The variation in FM_1 between bursts was minimal for frequencies in the range of 55–75 Hz. (Freeman 1985a)

3.2 Classification of Spatial Patterns

The mean density plots of sets of the five components were similar on visual inspection to the mean density plots of rms bursts amplitude for each rabbit. As previously described (Freeman and Schneider 1982, Viana Di Prisco and Freeman 1985) each rabbit had its unique spatial pattern of EEG amplitude, and the spatial patterns of the five components tended to conform to this pattern, although with greater variance for low energy components of individual bursts. An example of the patterns of the means of five components is shown in Fig. 7 for one rabbit; the control bursts and the CS+ and CS- T2, T3 bursts were averaged after deletion of those bursts having $f_1 < 55$ Hz.

The degree of similarity of the amplitude V_1 of dominant components within each subject and subgroup was evaluated by computing the product moment correlation coefficient r of each 1×64 matrix V_1 with other matrices V_1 and averaging by z-transform over subjects and burst pairs to find r. As expected, the average value r decreased monotonically with decreasing energy content of the components. Of greatest interest were the values for the domi-

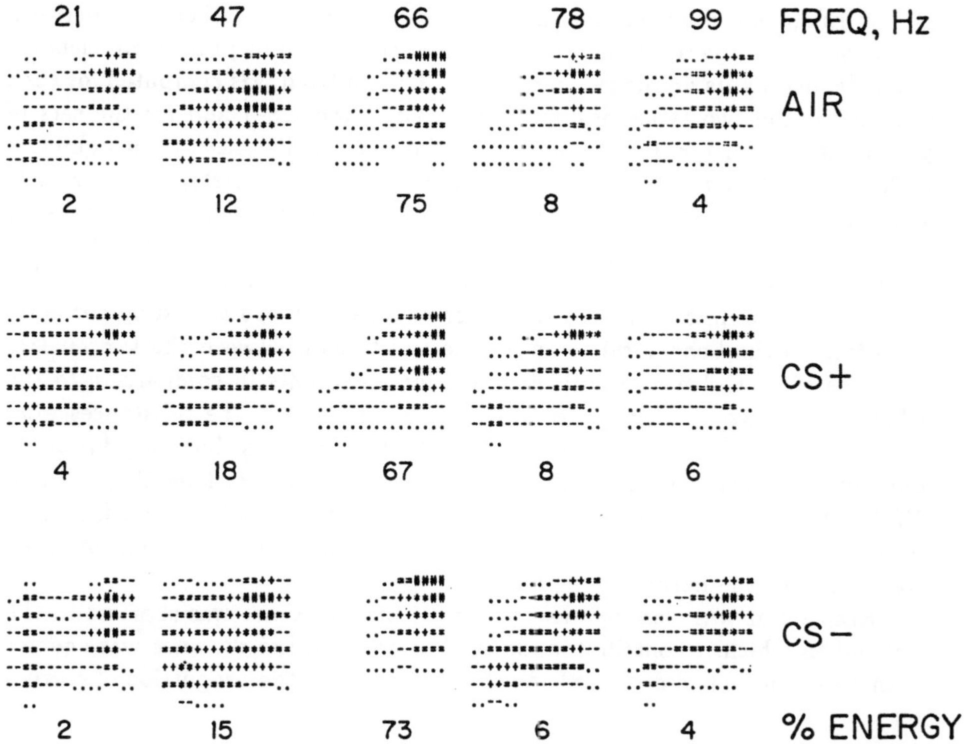

Fig. 7. Digital density plots were used to display 8 × 8 maps of the averages of the five components of control *CS+*, and *CS−* odor bursts (respectively *upper, middle* and *lower* rows for one subject) for those bursts having $f_1 < 55$ Hz. The mean of the frequencies is shown above each column; the proportion of relative energy expressed as% is shown under each frame. The spatial patterns of energy conformed in the main to the pattern of rms amplitude for each animal

nant components within the subsets of bursts with $f_1 > 55$ Hz. The dominant components of control bursts (based on sets of three bursts within trials) gave $\bar{r} = 0.97$, whereas this component for both CS+ and CS− bursts gave $\bar{r}_+ = 0.84$ and $\bar{r}_- = 0.88$ respectively, reflecting the lower amplitude of the odor bursts. The cross-correlation of control burst pairs C+ and C− preceding CS+ and CS− trials gave $\bar{r} = 0.85$, reflecting the introduction of between-trial variability. The comparison of the dominant components of CS+ versus CS− bursts gave $\bar{r}_{+-} = 0.58$, indicating that these spatial patterns across trials were substantially less similar to each other than were the control bursts across trials (Freeman 1985a).

The within-trial correlation of dominant components from bursts with $f_1 < 55$ Hz (those in which the lower frequencies tended to dominate) gave $\bar{r} = 0.12$, and similarly for the secondary and yet smaller components, and for dominant versus secondary, indicating that the components from these bursts did not form coherent classes within or over trials.

A Euclidean distance measure was used to evaluate differences between bursts with $f_1 > 55$ Hz. Each burst was treated as a point in a 64-dimensional space defined by its dominant component. The centroid was computed for each subject for the two types of odor bursts (CS+ and CS–) and for the sets of preceding control bursts (C+ and C–). The sets of distances from points to each centroid (their own or the centroids of other sets of bursts) were normally distributed. The standard error (SE) was computed for each of the four sets. The distance between each pair of centroids was divided by the pooled standard error of the two distributions to give a t-distance. Examples are given in Table 1 for the five rabbits ranked in the order of reliability of responding correctly according to the behavioral criteria. The t-distances between the two control centroids (C+, C–) were relatively small for all rabbits in comparison to the t-distances between control and CS+ or CS– centroids. The t-distances were relatively large also between the CS+ and CS– centroids, but only for those four rabbits showing behavioral evidence for discrimination. The (C+, C–) and (CS+, CS–) t-distances were greater for the subsets of trials on which correct responses occurred (as shown in Table 1) than for the entire sets of trials irrespective of the nature of the CR's.

Each burst was classified as "correct" if its t-distance was shortest to the centroid to which it contributed, or as "incorrect" if the distance was shorter to an apposing centroid (Fig. 8). As shown in Table 1 the proportion of correct

Table 1. *Upper left*: the numbers of *CR+* (lick) and *CR–* (sniff only) responses are shown for 30 *CS+* and 30 *CS–* trials. The conditioning procedure used established a lower rate for *CR–* than for *CR+*. *Upper right*: the Euclidean distance measures are shown between control (*C+*, *C–*) and odor (*CS+*, *CS–*) centroids, $f_1 > 55$ Hz. *Lower lef*: the proportions are shown of correct classification of bursts by the Euclidean distance measure. Chance level is 50% . *Lower right*: The results are summarized for the t-distance measure based on nonlinear mapping (see Fig. 9)

Number in 30 trials							
CR+		*CR–*		t-distance in 64-space			
CS+	*CS–*	*CS–*	*CS+*	air.air	air.*CS+*	air.*CS–*	*CS+*.*CS–*
25	2	18	5	2.58	5.78	3.87	4.97
25	2	10	3	2.98	7.69	12.55	4.87
20	9	13	5	1.28	11.27	4.27	5.08
19	3	10	9	2.87	3.21	4.74	4.20
4	4	7	23	2.01	10.41	14.60	1.22

Classification % correct				t-distance in nonlinear map			
Air.air	air.*CS+*	Air.*CS–*	*CS+*.*CS–*	Air.air	Air.*CS+*	Air.*CS–*	*CS+*.*SC–*
49	83	94	88	0.08	7.03	1.63	1.40
62	82	83	84	0.31	0.63	1.20	0.75
59	84	75	78	0.12	1.59	0.79	0.97
52	76	78	72	0.51	2.44	8.31	1.19
67	83	86	52	2.68	1.87	7.27	0.12

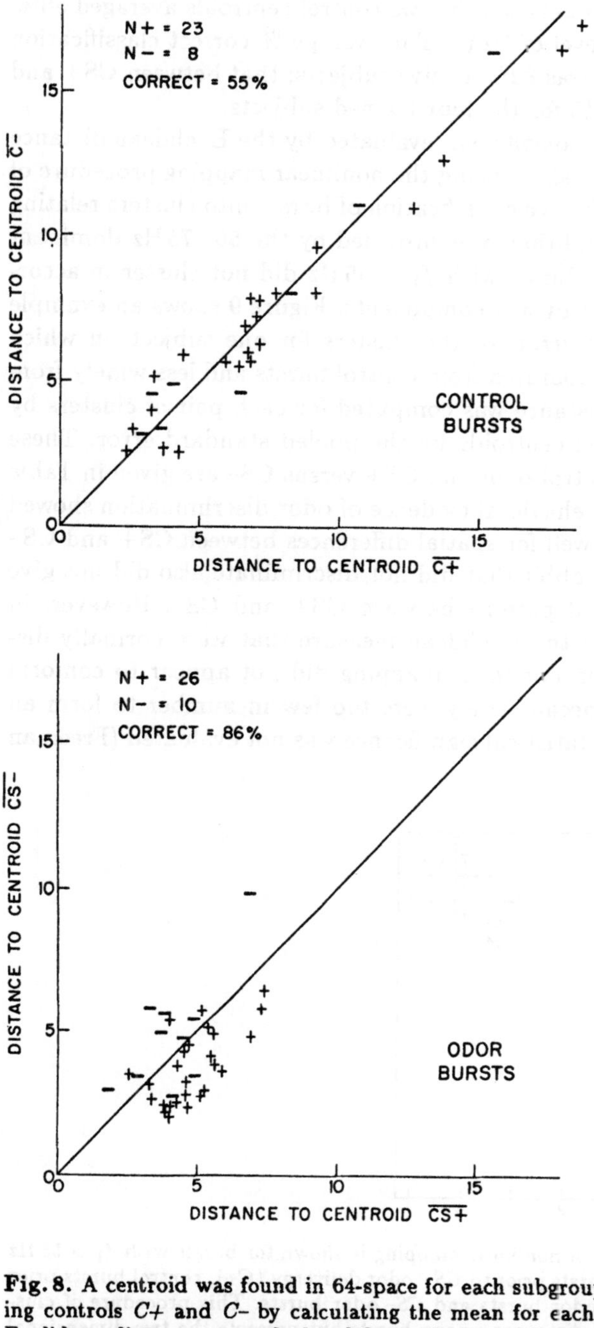

Fig. 8. A centroid was found in 64-space for each subgroup: *CS+*, *CS–*, and the two preceding controls *C+* and *C–* by calculating the mean for each channel over the subgroup. The Euclidean distance was calculated for each control burst of both types to the *C+* and *C–* centroids (*upper frame*), and for each odor burst of both types to the *CS+* and *CS–* (*lower frame*). If the burst point lay closest to its own centroid it was labelled "correct", otherwise "incorrect". The aim of measurement was to separate the odor bursts from each other (*CS+* from *CS–*) without separating the control bursts (*C+* from *C–*). The efficacy of various procedures was expressed by the difference between% correct odor bursts less% correct control bursts. (Freeman 1985a)

classification of control bursts between the two control centroids averaged 58%, which was near the chance level of 50% . The average % correct classification of odor versus control bursts was 82% for five subjects; that between CS+ and CS– odor bursts averaged 81% for the four trained subjects.

Each of the sets of 5 components was evaluated by the Euclidean distance measure and also by cluster analysis using the nonlinear mapping procedure of Sammon (1969). The only effective classification of bursts into clusters relating to control, CS+, and CS– conditions was provided by the 55–75 Hz dominant component with $f_1 > 55$ Hz. Bursts with $f_1 < 55$ Hz did not cluster in accordance with behavioral criteria by any components. Figure 9 shows an example of the controls and standard errors of the clusters for one subject, in which the odor bursts were widely separated from control bursts and less widely from each other. An empirical t-distance was computed for each pair of clusters by dividing the distance between centroids by the pooled standard error. These values for control-control, control-odor and CS+ versus CS– are given in Table 1. The four rabbits showing behavioral evidence of odor discrimination showed evidence by this criterion as well for spatial differences between CS+ and CS– burst patterns, whereas the rabbit that did not discriminate also did not give bursts that differed in spatial pattern between CS+ and CS–. However, in contrast to the distance from the Euclidean measure that were normally distributed, these distances from nonlinear mapping did not appear to conform to the normal distribution; because they were too few in number to form an empirical distribution, their statistical significance was not evaluated (Freeman 1985a).

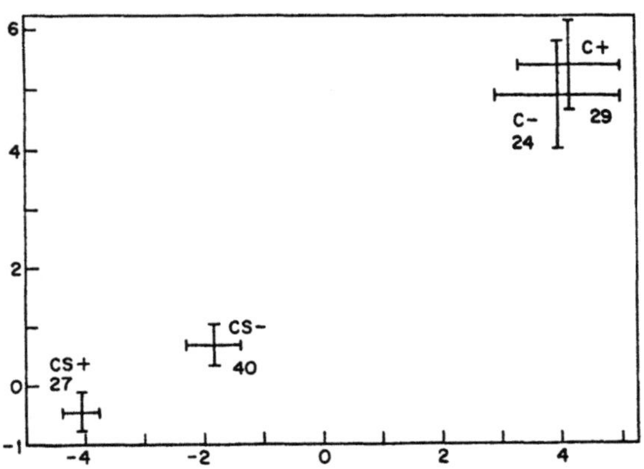

Fig. 9. An example of the results of nonlinear mapping is shown for bursts with $f_1 > 55$ Hz taken from four classes: control bursts prior to *CS+* odor delivergy (*C+*), control bursts prior to *CS–* odor delivery (*C–*), *CS+* odor bursts and *CS–* odor bursts. This procedure of cluster analysis preserves the relative distances among bursts but presents the two-dimensional display in a mode that minimizes the apparent scatter of those that are most similar to each other in spatial pattern. The results of further statistical analysis of the means of the centroids and their standard errors (*crosses*) are shown in Table 1. In all rabbits the means for controls *C+* and *C–* were not separated, whereas those for control and odor bursts were widely separated. In four of five rabbits the *CS+* and *CS–* clusters were separated; these rabbits also showed behavioral evidence for odor discrimination

The optimal cut-off level for f_1 was evaluated by repeatedly calculating the t-distances while raising the level in steps of 5 Hz beginning at 30 Hz and deleting bursts with f_1 below the level. Also deleted were bursts with $-50\% < FM_1 < 50\%$ and with $V_1 < 0.15$, both levels found by trial and error and accounting for 6% of bursts in this set. The procedure was done in two ways. One included only the C2, C3, T2, and T3 "discrimination" bursts on the premiss that these last two bursts were more likely to have occurred after discrimination had been achieved and thereby to give a clearer separation (Fig. 10, left frame). The other included all three test bursts and their control bursts to give a larger total number (right frame). Both sets included only trials on which a correct CR+ or CR− occurred, and only the 4 subjects that learned the CR+. Optimal separation of odor bursts with minimal separation of control bursts occurred in both ways with the cut-off at 55 Hz. A more sensitive measure was the % correct classification of odor bursts minus the % correct classification for control bursts. Figure 11 shows that this measure peaked with

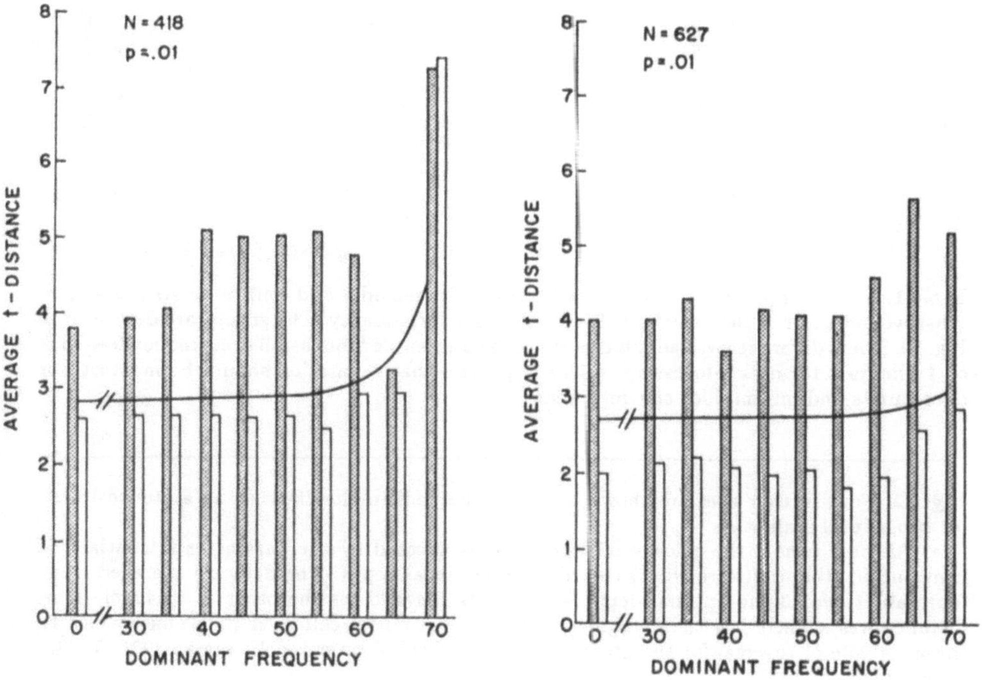

Fig. 10. Evaluation of the cut-off level for low frequency bursts was done by repeating the Euclidean distance measure after deleting the bursts below a cut-off level first at 30 Hz and then up to 70 Hz at increments of 5 Hz. The effects on the t-distances between $CS+$ and $CS-$ centroids (*shaded bars*) and between $C+$ and $C-$ centroids (*open bars*) are summarized for trials on which a correct CR occurred in four rabbits. The *left frame* includes only C2, C3, T2 and T3 bursts; the *right frame* includes all six bursts from each trial. The assignment of the p-value is based on the number of burst pairs remaining, on the normality of the distributions of the t-values, and on the assumption of independence among bursts. Optimal separation of odor bursts with minimal separation of control bursts was found at a cut-off of 55 Hz. (Freeman 1985a)

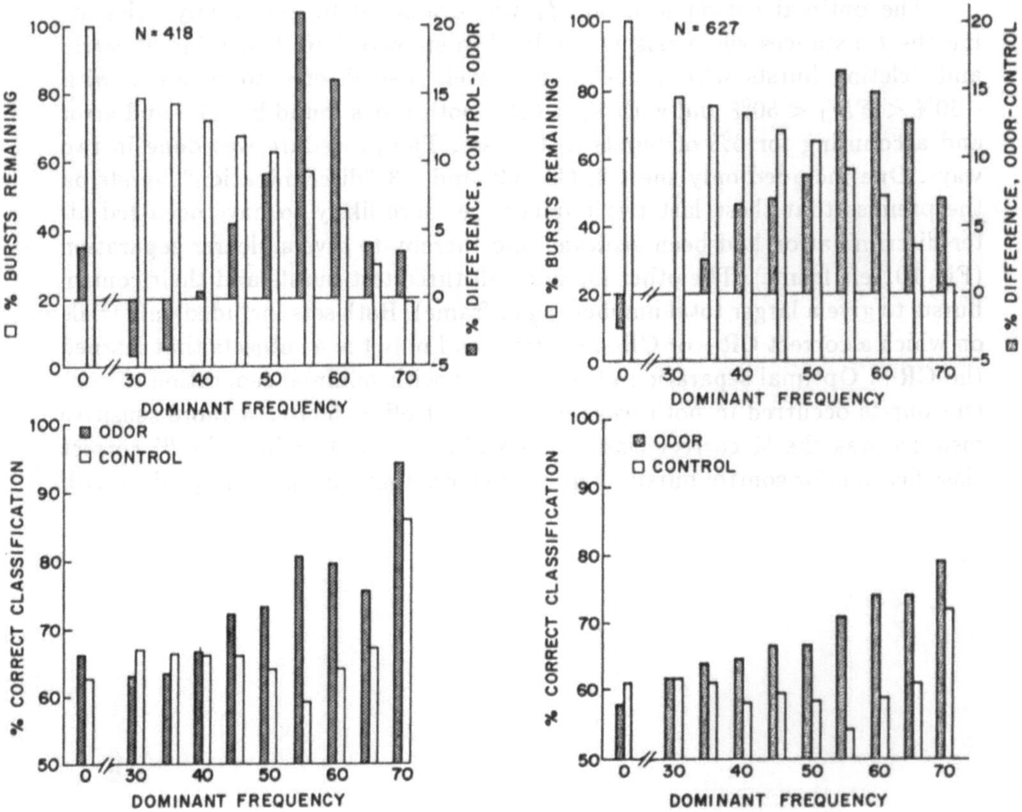

Fig. 11. The difference in % correct classification between odor and control bursts was a more sensitive indicator of the optimal value for the cut-off frequency. The groups are described in Fig. 10. The% difference was adopted as simple and versatile "touchstone" correction feedback on techniques for data processing, with the proviso that separation should be maximal for odor bursts and minimal for control bursts

Fig. 12. Four further uses are shown of the% difference in classification as a "touchstone" for procedural evaluation.

At *lower right* is the efficacy of separation as affected by the parameter in spatial deconvolution (the "software lens") that determines focal depth. The data are averaged from four rabbits; for 3 the optimal depth was at 0.49 mm and for the other at 0.54 mm. This distance was slightly less than the mean depth of the mitral cell layer (0.58 mm), which is the usual site of reversal for the granule cell dipole field of potential (Freeman 1975, 1980). From Freeman (1985b).

At *lower left* the "touchstone" is used to evaluate the effects of n-th order exponential spatial filters on the EEG (Gonzalez and Wintz 1977). The *upper curve* shows that maximal separation of odor bursts occurs with a 4th-order low-pass filter cut-off frequency (down 3 db) at 0.5 c mm^{-1}, as predicted by spatial spectral analysis of the bulbar EEG. The lower curve shows that minimal separation occurs with a 2nd order high-pass filter set at 0.17 c mm^{-1}, as predicted by the average of measurements of the widths of active EEG foci, and near 0.1 c mm^{-1} as predicted by the width of the electrode array. The order parameter (n) is optimized in the same way. These results prove that the odor-specific information exists in the spatial pass-band that is characteristic of the bulbar granule cell EEG generator (Freeman 1975, 1980).

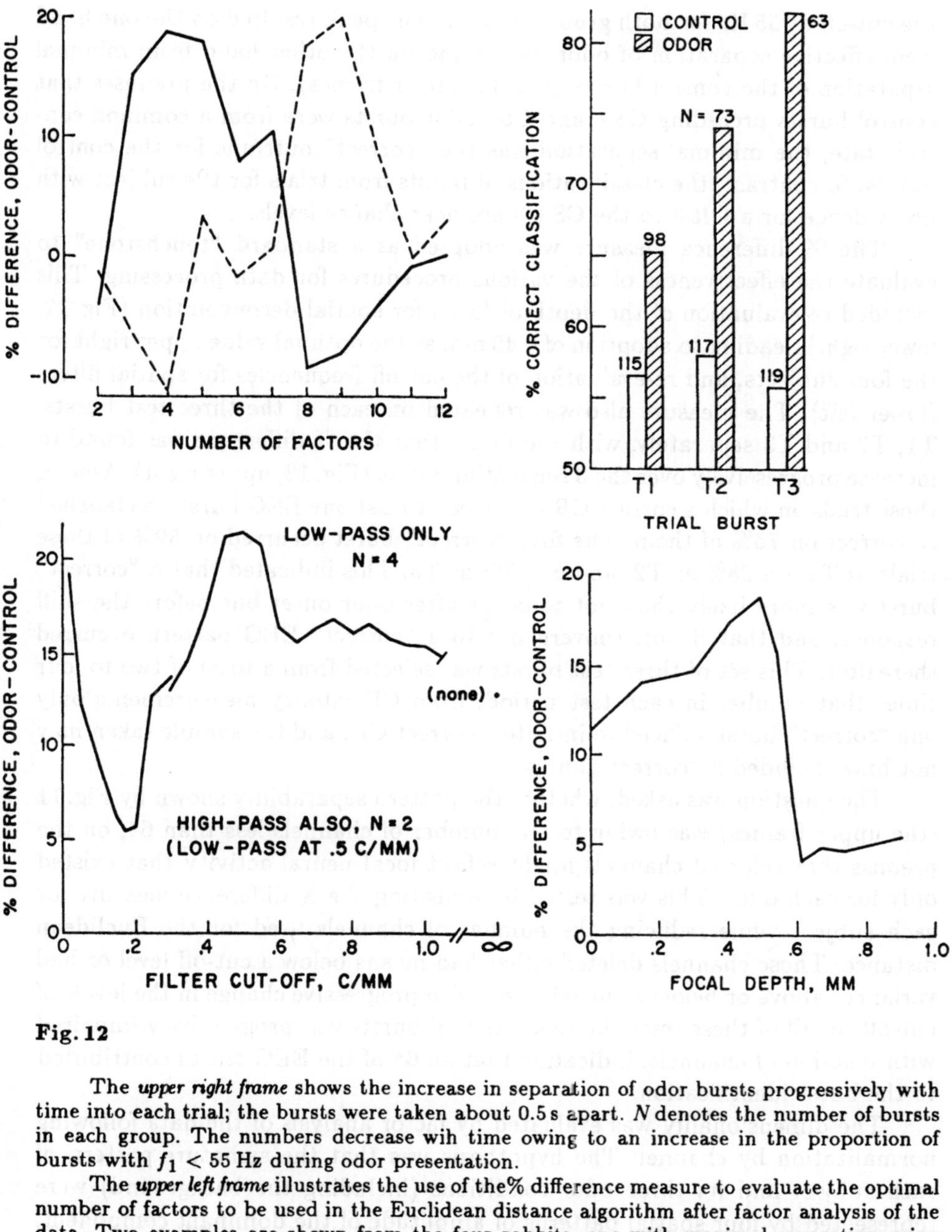

Fig. 12

The *upper right frame* shows the increase in separation of odor bursts progressively with time into each trial; the bursts were taken about 0.5 s apart. *N* denotes the number of bursts in each group. The numbers decrease wih time owing to an increase in the proportion of bursts with $f_1 < 55$ Hz during odor presentation.

The *upper left frame* illustrates the use of the% difference measure to evaluate the optimal number of factors to be used in the Euclidean distance algorithm after factor analysis of the data. Two examples are shown from the four rabbits, one with maximal separation using 4 factors, the other with 9 factors. When all factors with nonzero eigenvalues are used, the subgroups all approach 100% correct classification, and the% difference measure goes to zero. (From Freeman 1985c) the cut-off at 55 Hz for both groups of data. The peak resulted on the one hand from effective separation of odor bursts and on the other hand from minimal separation of the control bursts (Fig. 11, upper frames). On the premises that control bursts preceding CS+ and CS– odor bursts were from a common control state, the minimal separation was the "correct" outcome for the control bursts. In contrast, the classifications of bursts from trials for the subject with no evidence for a CR+ to the CS+ were near chance levels.

The % difference measure was adopted as a standard "touchstone" to evaluate the effectiveness of the various procedures for data processing. This included re-evaluation of the depth of focus for spatial deconvolution (Fig. 12, lower right), leading to adoption of 0.49 mm as the optimal value upper right for the four subjects, and re-evaluation of the cut-off frequencies for spatial filters (lower left). The measure also was repeated on each of the three test bursts, T1, T2 and T3 separately, with the result that the % difference was found to increase progressively over the 3 sequential bursts (Fig. 12, upper right). Among those trials on which a correct CR occurred at least one EEG burst was labelled as correct on 76% of them. The first "correct" burst occurred on 59% of these trials as T1, on 28% as T2, and on 13% as T3. This indicated that a "correct" burst was more likely than not to occur after odor onset but before the sniff response, and that if not, convergence to a "correct" EEG pattern occurred thereafter. This set of three test bursts was selected from a total of two to four times that number in each test period; from CR latency measurements only one "correct" burst sufficed to initiate a correct CR, and the sample taken may not have included a "correct" burst.

The question was asked, whether the pattern separability shown by Fig. 11 (the upper frames) was owing to any number of channels less than 64, on the premiss that selected channels might reflect focal neural activity that existed only for each odor. This was tested by repeating the % difference measure for each subject while reducing the number of channels used for the Euclidean distance. Those channels deleted either had means below a cut-off level or had variances above or below a cut-off level, with progressive change in the levels of cut-off. In all of these tests the separation of bursts was progressively impaired with deletion of channels, indicating that all 64 of the EEG traces contributed to the burst identification.

The dimensionality was evaluated by factor analysis of the data following normalization by channel. The hypothesis was that the signature pattern of each subject and its three odor conditions (including the background) were represented by four spatial patterns of amplitude of the dominant component, and that these patterns might be accessed through the sets of eigenvectors from the correlation matrix. The data set for each subject was re-expressed in this new coordinate system of the factor space; the channel amplitudes were replaced by factor scores. These scores were used in the Euclidean distance measure to determine the optimal number of factors to estimate the degree of separation. Two examples are shown in Fig. 12 (upper left) of the change in the % difference measure as the number of factors was increased from 2 to 12. In two subjects the optimal number of factors was 4; in two others it was 8 or 9; these subsets of factors incorporated 92% of the variance. The average % difference of 24.2% (80.0% – 55.8%) did not differ from the value of 20.8% (80.2% – 59.4%) from analysis of channels (Fig. 11, left frame). Under varimax rotation the optimal number factors ranged from 3 to 5 over the four subjects with the % difference at 24.8% (78.4% – 53.6%); under oblique rotations the % difference deteriorated, indicating that the factors indeed were orthogonal. In this way the number of variables used for classification of bursts (the 3 to 5 factors) was less than the number of bursts (ranging from 8 to 42 and averaging 17) in every classification subgroup (C+, C–, CS+, and CS–), on the average about four fold.

Three tests were made for factorial invariance. In the first test factor analysis was repeated n times over n cases for each subject, with deletion of a different burst on each run ("jack-knifing"). The grand mean for the % difference was 20.6% (73.9% – 53.3%). The standard deviations about the subject means (ranging from –1.4% to 38.2%) ranged from 2.98% to 5.63% with an average value of 3.78% . The distribution of % difference values combined after normalization by subject was not significantly different from normal by the chi square test, indicating that the 99% confidence interval for the % difference was +/–8.80% . By this criterion the % difference values reflected differences between CS+ and CS– bursts that were significant for each subject at $p < 0.01$.

In the second test factor analysis was repeated on subsets of the data, and the similarity of the factor patterns was determined from correlation matrices of the factors. Table 2A shows the matrix resulting from factor analyses of the even- and odd-numbered bursts for each subject. The correlation coefficients were averaged over subjects by Fisher's z-transform, after reranking factors in three instances when their order by size of the eigenvalues had changed. The diagonal entries for the first four factors averaged 0.904, showing that both subsets of the data supported convergence to a common set of spatial patterns. Table 2B shows the results from performing factor analysis on three sets of bursts, one from each of the three sessions at weekly intervals that provided the whole data set, and averaging over subjects and session pairs. The high values on the diagonal again reflected the stability of the factor patterns extracted from the EEG over a period of weeks.

In the third test factor analyses were done independently on the odd- and even-numbered bursts respectively as "training" and "validation" sets. In this test the entire set of trials was used, including those with and without correct CR's, in order to increase the number of cases; this reduced the average % difference to 17.9% (70.5% – 52.6%). Factor scores were determined for all bursts with respect to both sets of factors. The lower left half of Table 3 shows the % difference measure of classification with factor scores of the same subsets of bursts as used to derive the factors (the "training" sets); the lower right half shows the % difference measure with factor scores of the omitted subsets (the

Table 2. The results are shown of using correlation over factors derived from analysis of subsets of the data to test for factorial invariance. In Table 2A the data were divided into two subsets by partitioning the entire data set into two parts, one of odd-numbered bursts and one of even-numbered bursts. In Table 2B the factor analysis was done on the data from each of the consecutive three behavioral sessions yielding the bursts. The asterisks denote the correlation coefficients of the first four factors (re-ranked in the few instances where the order by rank of the eigenvalues differed over sessions) averaged (by Fisher's *z*-transform) over subjects and (in Table 2B) inter-session comparison. These high coefficients reflect the stability of the factor patterns over sessions. From Freeman (1985c)

A

Factors	I	II	III	IV	V	VI
I'	0.94*	-0.23	-0.31	0.02	-0.05	-0.07
II'	-0.08	0.95*	-0.29	-0.32	-0.14	0.10
III'	-0.31	-0.21	0.86*	0.04	-0.11	0.02
IV'	-0.10	-0.37	-0.18	0.80*	-0.09	0.02
V'	-0.21	-0.11	0.01	0.09	0.41	0.21
VI'	0.06	-0.13	-0.03	0.04	0.21	0.33

B

Factors	I	II	III	IV	V	VI
I'	0.86*	-0.09	-0.28	-0.02	0.00	0.01
II'	-0.19	0.94*	-0.35	-0.38	-0.19	-0.13
III'	-0.34	-0.20	0.79*	-0.15	-0.17	0.04
IV'	0.06	-0.49	0.07	0.83*	-0.08	-0.00
V'	-0.17	-0.13	0.06	0.14	0.46	0.22
IV'	0.07	-0.14	-0.06	0.04	0.17	0.39

Table 3. The bursts from all trials (whether or not a *CR* occurred) are divided into those odd- and even-numbered. Classification by the% difference measure on four factors is as effective (within the precision of the measure determined by "jack-knifing") for "validation" subsets (*lower right*) as for "training" subsets (*lower left*). Subject 5 showed no behavioral evidence for odor discrimination. From Freeman (1985c)

Numbers of bursts in separated groups		Odd				Even			
Subject	Total	*C+*	*CS+*	*C−*	*CS−*	*C+*	*CS+*	*C−*	*CS−*
1	86	13	7	10	13	9	11	13	10
2	80	6	19	5	10	9	16	6	9
3	120	9	19	12	20	19	10	14	17
4	92	13	10	11	12	7	14	12	13
5	120	17	13	16	14	8	22	10	20

% difference classification measure	Odd*odd	Even*even	Odd*even	Even*odd
	$X_1 \times X_1$	$X_2 \times X_2$	$X_1 \times X_2$	$X_2 \times X_1$
1	26.2	18.6	26.2	28.1
2	- 5.2	18.2	- .52	26.6
3	9.2	16.1	9.2	18.1
4	15.2	4.3	19.0	12.3
Avg.	11.4	14.3	12.3	21.3
5	- 4.3	- 8.8	- 1.3	- .27

"validation" sets). The results showed that the factors derived from one half of the data were no less effective in classifying the other half, the differences being within the error of measurement as determined by "jack-knifing".

Discriminant analysis was done on the factor scores after Varimax rotation. The two sets of C+ and C– control bursts were pooled, and the question was asked whether by two discriminant functions the three groups of control, CS+, and CS– bursts were from a common population. This case was rejected by Wilks lambda criterion at $p = 0.01$ for all four rabbits, with F values ranging from 2.69 (d.f. 8,108) to 15.79 (d.f. 8,202). The over-all level of correct classification of the individual bursts with respect to the three centroids by a Euclidean distance measure (Cooley and Lohnes 1962) was over 80% for the four rabbits.

4 Discussion

These results show the EEG bursts (optimally about 80%) can be sorted into classes relating to the control state and the two odor states on the basis of the spatial pattern of amplitude of the dominant cosine component. The separation holds for bursts from trials with a correct CR+ or CR–; on incorrect and nonresponding trials the separation approaches chance levels. On the majority of correct trials the separation begins with the first burst after CS onset, and it improves progressively through the third burst. The separation of CS+ and CS– odor bursts holds only for the four of five rabbits that learned the CR+, although the fifth rabbit that did sniff to both odors shows a significant difference between control pattern and the common odor spatial pattern. On these grounds we conclude that the EEG manifests odor-specific information, and that this information takes the form of stable and reproducible spatial patterns embedded in the neural activity accompanying the odor and the control states both within and across the sessions having stable behavior.

These results are based on the exclusion from factor and cluster analysis of bursts having a dominant frequency $f_1 < 55$ Hz. The validity of our interpretation hinges on a physiological explanation of the origin and significance of these as aberrant bursts. We characterize them as having frequencies spread relatively widely but usually in the 20–50 Hz range; a tendency to augmenting frequency within each burst; a large variation in the amount of frequency modulation from each burst to the next; a relatively low fraction of total burst energy in the dominant component; and a spatial amplitude pattern that varies unpredictably, in contrast to the reproducible spatial patterns of bursts with $f_1 > 55$ Hz. All of these features indicate that bursts with $f_1 < 55$ Hz are less coherent than those with $f_1 > 55$ Hz.

In these respects they resemble the background or interburst activity. This so-called "spontaneous" activity is most clearly seen in waking, unmotivated animals in which bursts seldom occur. It has the appearance of random activity in several respects; its time series is irregular and unpredictable; its spectrum has multiple low peaks that vary from one segment to the next; its amplitude histogram is Gaussian; and the envelope of the time series conforms to the Rayleigh distribution (Freeman 1975). Remarkably, however, in every time segment the same seemingly random time series appears on all electrodes in the array, and the ensemble average of the spatial rms amplitude over multiple interburst segments has the same form as the unique signature pattern for each rabbit, which we derive from ensemble averages of dominant components from the coherent and the incoherent bursts.

We conclude that this activity is not random. On the one hand, there is no known random generator located in the bulb that can serve as a noise source to produce this temporally widespread incoherent activity with a common instantaneous frequency at all times. Yet the source must lie within the bulb, because (a) the receptors do not have the physiological basis for the requisite coordination of their activity, other than that imposed at low frequency (3–5 Hz) by inhalation, and (b) the activity persists after transection of the connections of the bulb to the brain. On the other hand, we know that the bulb can be regarded as a distributed system of coupled oscillators. From simulation studies we have the capability of generating chaotic activity with such a system (Freeman 1979), as well as both steady state and limit cycle activity patterns. We conclude that the background state is the manifestation of a chaotic attractor of the bulbar mechanism and not of an equilibrium or point attractor under perturbation. We infer that the low-level basal and interburst activities are manifestations of a bulbar chaotic attractor, and that the great majority if not all of bursts with $f_1 < 55$ Hz are manifestations of one or more chaotic attractors, not that of the basal state (its peak frequency exceeds 55 Hz) but another one or more with properties similar to it.

A stable and reproducible spatial pattern of coherent activity repeatedly emerges from chaos and is manifested in the dominant component of the control bursts in each rabbit with $f_1 > 55$ HZ; this indicates that a limit cycle attractor exists in the mechanism, to which the output tends to converge upon most inhalations. The discovery of differing stable spatial patterns during inhalation of CS+ and CS– odors implies that a limit cycle attractor may exist for each discriminable odor.

We suggest that the changes in spatial EEG pattern that occur during learning to respond to an odor (Freeman and Schneider 1982, Viana Di Prisco and Freeman 1985, Gray et al. 1985) reflect a process of synaptic modification and nerve cell assembly formation (Freeman 1979, 1981) by which a new limit cycle attractor is formed. Looking toward the receptors, a basin is formed such that, under the initial conditions set by the input during learning, the bulbar mechanism moves toward one or another differing limit cycle state; looking toward the olfactory cortex, the output of the bulb is a reproducible spatial pattern of activity that is determined by the attractor of the basin selected by the input. We suggest that the control burst pattern is determined by learning in respect to the collection of background odors, and that the background, CS+

and CS– spatial patterns represent three of a repertoire which includes other undefined limit cycle attractors.

By this interpretation each inhalation constitutes a test of the environment by the rabbit, that may result in convergence of the bulbar mechanism to a limit cycle attractor, but that may result on some inhalations in failure of convergence with an indeterminate outcome. We speculate that this event may be manifested in those bursts in which the energy is shifted to a lower frequency manifesting a chaotic attractor. This might provide the basis for a clear signal to the rabbit of indeterminacy, leading to re-testing or to initiation of an orienting response.

The spatial pattern relating to each CS could not be localized to any smaller number of channels than the set of 64. When channels were deleted by any of several criteria relating to amplitude, phase, variance, or change from control, the efficacy of separation by the % difference measure and by nonlinear mapping diminished. This did not result simply from reducing the dimensionality of the data sets, as shown by factor analysis. This result indicates that the spatial pattern of each attractor exists widely over the bulb, perhaps the whole of it (Bressler 1984), that our electrode array gives us only a spatial sample, and that each part of the sample contributes to the resolution of pattern differences. In the manner of analysis of a large scene, parts may be deleted without loss of the global image but with diminished resolution from other patterns.

These results leave questions unanswered regarding the size, internal structure, and fineness of spatial grain of the pattern of a limit cycle attractor. The array window covers less than 20% of the bulbar surface, and the interelectrode distance (0.5 mm) is twice the mean interglomerular distance. Four to 8 times the present number of electrodes would improve spatial resolution by increasing the window size and decreasing the interelectrode distance, but at great experimental cost if temporal resolution were not to be sacrified. At present the centroids of clusters are clearly separable, but the percentage of correct classification of individual CS+ and CS– bursts based on the results of the distance measure averages 80% . This is near the average 74% of correct behavioral responses by discriminating rabbits for the CS+, but there is not a one-to-one correspondence between correctness of classification by the behavioral and EEG measures over the set of 180 odor bursts from each rabbit. Improved pattern resolution is needed, including more extensive sampling of the EEG throughout the interval between CS and CR. The manner of read-out of limit cycle and chaotic attractor patterns is undetermined. This should depend on spatial as well as temporal integration by the cortex over the bulbar output, but the kernels of the integrals and the cortical operations on bulbar output are as yet unknown.

The KII model (Freeman 1979) that has been used to simulate the temporal and spatial patterns of EEG waves has not been adequately explored for the properties of compound or multiply co-active templates to simulate nerve cell assemblies. Regarding chaos these new findings should be simulated with coupled nonlinear differential equations, in the manner that we have simulated the onset and termination of bursts in the form of limit cycle activity. Our qualitative analysis of chaotic activity must be verified by quantitative analysis, including estimation of the fractal dimensionality of the putative chaotic attractors (Freeman 1985d).

The results explain why the analysis of burst root mean square (rms) amplitude patterns (Viana Di Prisco and Freeman 1985) failed to detect reliable differences between CS+ and CS− bursts. First, the differences are not clear-cut unless the patterns are spatially deconvolved by the "software lens" (Freeman 1980). This operation is very sensitive to noise and cannot be used effectively unless the noise is attenuated with both spatial and temporal filters. Further, spatial filtering cannot at present be done prior to Fourier decomposition of the time series at each channel. Second, the bursts with $f_1 < 55$ Hz, which we propose to call chaotic, constitute a massive source of variance. When all bursts are pooled, an essential condition for factor analysis is violated, and the results lead to excessive scatter of the classes of CS burst patterns. The results also explain how the 64 channel array can suffice to detect burst pattern differences. If the spatial pattern of the limit cycle attractor is widely distributed in the bulb, then the precision of anatomical placement of electrodes is secondary, and the degree of resolution depends mainly on the number of available electrodes.

We conclude that the bulb operates as a global dynamic system in odor discrimination (Freeman 1985c, Freeman and Skarda 1985). Its background activity is determined by a chaotic attractor (Freeman 1985d) that allows for easy transition to other attractors. Input during inhalation determines the basin for another attractor and forces a bifurcation by an attendant parametric increase in feedback gain (Freeman 1979). For each odor discriminated there is a learned, limit cycle attractor, that determines a global spatial burst pattern signifying the presence of that odor. If convergence fails, or if there is no attractor corresponding to the input odor, then a chaotic burst occurs, typically at a lower frequency, signifying the need for re-testing, orienting, habituation or learning depending on reinforcement (Gray et al. 1985). This allows for updating of the mechanism in an ever changing environment.

References

Abraham RH, Shaw CD (1983) Dynamics-The geometry of behavior. Aerial Press, Santa Cruz

Bressler SL (1984) Spatial organization of EEGs from olfactory bulb and cortex. Electroencephalogr Clin Neurophysiol 57:270–276

Cooley WW, Lohnes PR (1962) Multivariate procedures for the behavioral sciences. Wiley, New York

Freeman WJ (1975) Mass action in the nervous system. Academic, London New York

Freeman WJ (1978) Spatial frequency analysis of an EEG event in the olfactory bulb. In: Otto DA (ed) Multidisciplinary perspectives in event-related brain potential research, pp 531–546. US Govt Printing Office, EPA-600/9-77-043

Freeman WJ (1979) EEG analysis gives model of neuronal template-matching mechanism for sensory search with olfactory bulb. Biol Cybern 35:221–234

Freeman WJ (1980) Use of spatial deconvolution to compensate for distortion of EEG by volume conduction. IEEE Trans Biomed Eng 27:421–429

Freeman WJ (1981) A physiological hypothesis of perception. Perspect Biol Med 24:561–592

Freeman WJ (1985a) Relation of olfactory EEG to behavior: Time series analysis. Behav Neurosci (in press)

Freeman WJ (1985b) Relation of olfactory EEG to behavior: Spatial analysis. In preparation

Freeman WJ (1985c) Relation of olfactory EEG to behavior: Factor analysis. In preparation

Freeman WJ (1985d) Analytic techniques used in the search for the physiological basis of the EEG. In: Gevins A, Remond A (eds) Handbook of electroencephalography and clinical neurophysiology, vol III A, part 2, chap 14. Elsevier, Amsterdam (in press)

Freeman WJ, Schneider W (1982) Changes in spatial patterns of rabbit olfactory EEG with conditioning to odors. Psychophysiology 19:44–56

Freeman WJ, Skarda CA (1985) Nonlinear neural dynamics, EEG, and perception: the neo-sherringtonian view. Brain Res Rev (in press)

Garfinkel A (1983) A mathematics for physiology. Am J Physiol 14:R455–R466

Gonzalez RC, Wintz P (1977) Digital image processing. Addison-Wesley, Reading

Grassberger P, Procaccia I (1983) Measuring the strangeness of strange attractors. Physica 9D:189–208

Gray CM, Freeman WJ, Skinner JE (1985) Associative changes in the spatial amplitude patterns of rabbit olfactory EEG are norepinephrine dependent. In preparation

Sammon JW (1969) A nonlinear mapping for data structure analysis. IEEE Trans Comput 18:401–409

Viana Di Prisco G, Freeman WJ (1985) Odor-related bulbar EEG spatial pattern analysis during appetitive conditioning in rabbits. Behav Neurosci 99:964–978

Tensor Network Theory of the Central Nervous System and Sensorimotor Modeling

A.J. PELLIONISZ[1]

1 Brain Theory as a Unifying Core for Disciplines Linked by Cybernetics

Brain theory is quickly developing into a central focus among disciplines as varied as Neuroscience (Reichardt and Poggio 1981), Machine Intelligence and Robotics (Marr 1982, Palm 1982, Loeb 1983), Computer Science and Mathematics (Grimson 1981), Health Care and Rehabilitation (Mann 1981), and Philosophy and Humanities (Churchland 1985). The unifying power of brain theory, however, is not a recent discovery: the assumption that *all* brain functions can be more coherently represented once the functioning of the brain itself is basically understood, is as old as brain theory. The desire for a unified view goes back to Plato's idea that the external world is reflected by forming a single image in the darkness of a cave.

For centuries, however, the creation of a general framework to represent sciences and humanities in a global manner was not a problem of basic research, but a question of philosophy. Recently, an interface between philosophy and the neurosciences, called "neurophilosophy" (Churchland 1986) has been proposed to explain how such monistic representation, "a high level supersoftware" may be manifest in a *single brain* (cf. Llinàs and Pellionisz 1984). Philosophy aside, it was not until a generation ago that such unification was explicitly attempted at the level of *basic science* by establishing Cybernetics (Wiener 1948) as a novel, non-traditional interdisciplinary field of research.

Cybernetics, however, was not created to be an Archimedean fulcrum, a singularly solid pivotal point; rather, it was bi-centered around "control" and "communication". It served the role of a blanket, which covered a mosaic of disciplines instead of providing a new discipline, which would unify all of them. It has always been implicit, however, that with enough time and much effort, a true discipline would transpire from the conglomerate of Cybernetics – a traditional one that stands on its own footing, not on the fringes of many others. It is likely today, e.g. from this Vol., that such a metamorphosis from

[1] Department of Physiology and Biophysics, New York University, 550 First Avenue, New York, NY 10016, USA

Brain Theory
Edited by G. Palm and A. Aertsen

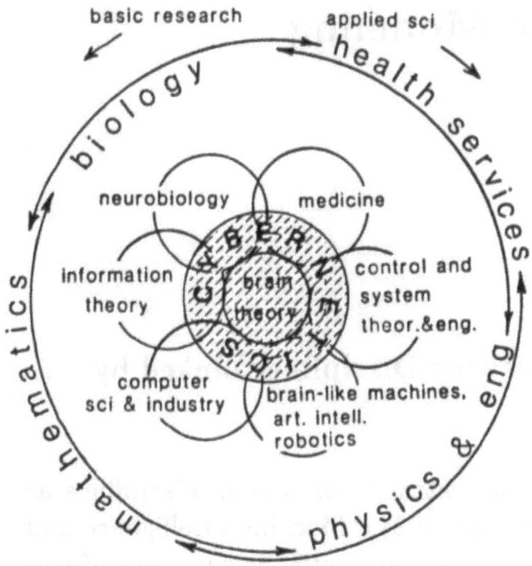

Fig. 1 Brain theory as the core for Cybernetics. Cybernetics was created as an interdisciplinary field of study to tie together various disciplines. Bi-centered around "communication and control", it lacked a conceptually and formally homogeneous theoretical core, which could be provided by brain theory

the interdisciplinary to the disciplined, will crystallize around a core provided by brain theory (see Fig. 1).

The establishment of such a solid pivotal point, however, requires a mathematically coherent explanation of how brain function emerges from neuronal networks in the central nervous system (CNS). Once a central concept matched with a powerful-enough formalism is developed into an accepted brain theory, a transfer of the understanding of CNS function into its utilizations, most particularly in medicine and in the science and technology of constructing brain-like machines (cf. Loeb 1983, Pellionisz 1983), will be possible.

Until this goal is reached, brain theory and neuronal modeling will exist as a plethora of divergent "system" and "subsystem" theories of the CNS and will continue to be relegated to auxiliary roles within and outside of the disciplines which were loosely covered by the umbrella of cybernetics. However, current trends indicate that brain theory will develop into a leading topic in the exact life science of biophysics, constituting its subdiscipline "neurophysics", since a measure of rigor in science is the extent to which mathematical theory is utilized. This general progress towards exact mathematical synthesis in neuroscience has been evident since the efforts of Helmholtz (1896) and more recently, of Hodgkin and Huxley (1952). Today, theory is a means of predictive modeling, and more fundamentally, a guide for economical and convergent research. The need for such efforts is underscored by statements like "There has been a tremendous explosion of data over the past decade or so but there is a surprising absence of theory to bring it all together" (Lewin 1982). Therefore, a search for general organizing principles of the CNS, as attempted in the rest of this paper, may prove to be essential.

2 Tensor Analysis Among Other Theoretical Approaches to Brain Function

Modern theoretical approaches to CNS function increasingly adopt the view that the brain is a distributed parallel processor (cf. Hinton and Anderson 1981). The acceptance of this principle has been difficult, in part, because this global view calls for experimental techniques with which to obtain multiple sets of data to adequately characterize the functioning of a parallel organization. Pioneering studies to investigate mass-action in the CNS have been conducted more than a decade ago (Bantli 1972, Freeman 1975, Llinàs et al. 1973). However, because of the difficulties inherent in such techniques, only recently have efforts reached very significant dimensions (cf. Gernstein et al. 1983, Bower and Llinàs 1983, Epping et al. 1984).

Beyond experimental techniques, the acceptance of the principle of parallel organization of the brain was made even more difficult because this view calls for adequate conceptual and formal representations of such parallel systems. Sherringon's original vision of "patterns on the enchanted loom" (1906) has been concretized since by notions of synergies (Bernstein 1947), cell assemblies (Hebb 1949) and patterns (Beurle 1962), which have been utilized even in recent times (Eccles 1973, Katchalsky et al. 1974, Palm 1982). From a structural point of view, however, the notion of morphological "modules" has gained ground (Mountcastle 1979, Eccles 1982). On the other hand, from a functional viewpoint, the activities of populations of neurons were increasingly described by vectors (Pitts and McCulloch 1947, Kohonen 1972, Anderson 1983, Hinton and Anderson 1981). Conventional representations usually assume, however, that the vectors are expressed in *extrinsic*, orthogonal systems of coordinates and that they represent points in ordinary Euclidean vectorspace.

2.1 The Axiom Underlying the Tensorial Approach: CNS Utilizes Intrinsic Coordinates

Tensor network theory of the CNS abandons the above assumption; it is physically evident that the experimentally revealed frames, *intrinsic* to the CNS, are in fact non-orthogonal (see Fig. 2, cf. Helmholtz 1896, Blanks et al. 1972, 1975, Simpson 1983, Simpson and Graf 1985, Simpson et al. 1981, Ezure and Graf 1984, Baker et al. 1985).

Building a novel scientific theory by replacing existing dogma with a new set of axioms is significant because such changes typically result in new possibilities for scientific development. The formally subtle yet conceptually axiomatic shift in the generally utilized vectorial approach – from describing CNS function in extrinsic frames to understanding CNS function in intrinsic coordinates – opens new avenues for brain theory. The standard, ubiquitous usage of the vector-matrix approach in neuroscience may be substantially improved by tying it to a general underlying concept (multidimensional geomery of brain function) and by further generalizing the abstract formalism (from coordinate-system-specific vector analysis to absolute differential calculus, expressible in any coordinate-system by general tensor notation). Some possibilities, immediately explorable by tensor theory which is based on the axiom of intrinsic, general systems of coordinates in the CNS, are shown below.

Mathematically, the CNS can be represented as expressing physical objects such as movements, in a multicomponent vectorial forms. Tensor analysis generalizes this point by stating that the CNS uses various multidimensional non-orthogonal frames that are usually different from one another and transforming vectors (a) from one frame to another (b) within one frame, from covariant-to-contravariant expression, via a metric tensor (Pellionisz and Llinàs 1980).

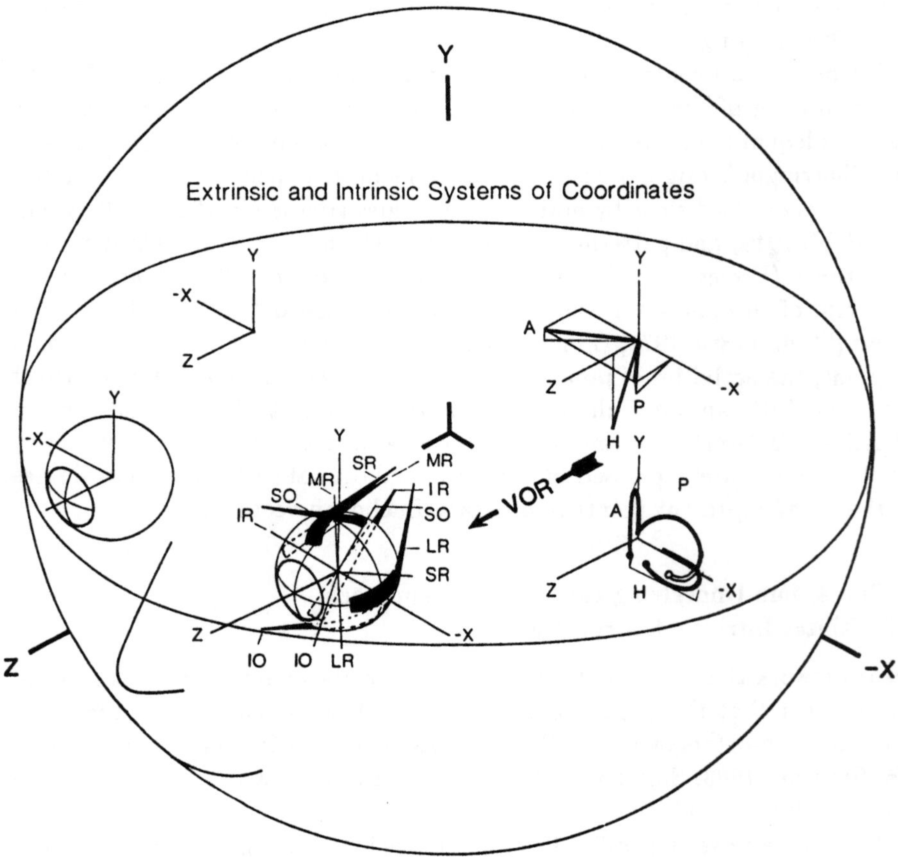

Fig. 2 An example of intrinsic coordinates in the CNS: Natural frames of reference for the vestibulo-oculomotor sensorimotor system (after Pellionisz 1985b). The vestibular semicircular canals (*A* anterior; *P* posterior; *H* horizontal) can be characterized by their rotational axes. For the motor system, the rotational axes of the eye correspond to the pull of extraocular muscles; *MR* medial rectus; *LR* lateral rectus; *SR* superior rectus; *IR* inferior rectus; *SO* superior oblique; *IO* inferior oblique. These rotational axes of the vestibular semicircular canals and the extraocular muscles constitute two built-in (intrinsic) frames of reference for the CNS to measure a head movement and to execute a compensatory gaze-shift. Such intrinsic systems of coordinates are typically non-orthogonal and often overcomplete general frames of reference

These known but largely overlooked expressions and conversions, schematically shown in Fig. 3 and discussed in further detail in Sect. 2.2, can be described by tensor transformations.

Biologically, transformations in general frames (tensor transformations) can be seen as implemented by particular connectivity-matrices of neuronal networks. Thus, a network theory is offered by the tensorial approach. Further, the distinction of covariant and contravariant expressions of a physical object even in one frame of reference provides a mathematical basis for an interpretation of sensorimotor integration. In a primary sense, sensory mechanisms operate with covariants, motor systems with contravariants. Thus, a sensorimotor theory is offered by the tensorial approach. Presently few other formal theories are capable of expressing sensorimotor integration both at a highly abstract conceptual level and at a quantitatively detailed network level.

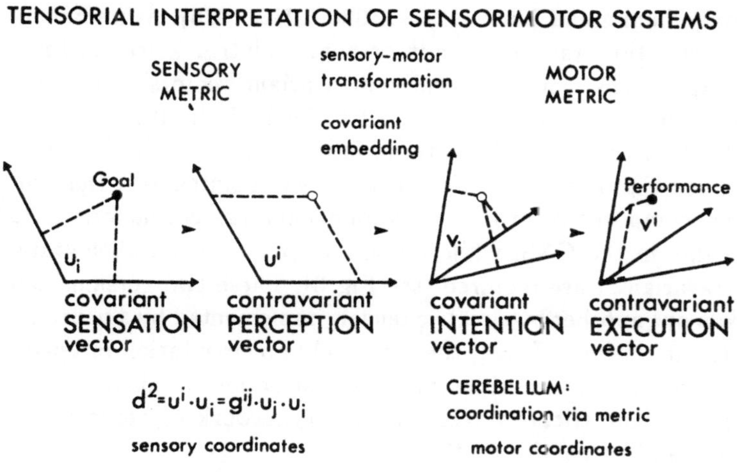

Fig. 3. The axiom that brain function is expressed in natural intrinsic frames of reference leads to a tensor theory of transformations in the CNS (via neuronal networks) among different vectorial expressions of physical invariants (after Pellionisz and Llinàs 1981). Sensorimotor systems express physical objects (e.g. a goal-oriented movement) in sensory and motor frames of reference (*left two* and *right two diagrams*, respectively). The frames may well be different from one another and usually are different from orthogonal Cartesian systems of coordinates. In the two general (non-orthogonal) frames, four different vectors can be rendered to a physical invariant. The two sensory expressions are the orthogonal projection-type (covariant) sensory reception vector with independently established components (SENSATION), and the parallelogram-type (contravariant) sensory PERCEPTION vector with interdependent components. The latter can be obtained from the former by a metric-tensor-type transformation-matrix, implementable by a neuronal network. With both sensory expressions available, their dot-product provides the invariant *d* distance (between the origin and the goal-points). A change of systems of coordinates from sensory to motor may be implemented by a covariant embedding transformation. This, however, yields an intention-type (covariant) motor vector, whose components physically do not synthesize the goal of the invariant. The executable contravariant physical (parallelogram) motor components that yield the proper performance can be obtained by an INTENTION-EXECUTION (covariant to contravariant) transformation, suggested to be implemented by the cerebellar neuronal network

2.2 Fundamental Notions of Tensor Network Theory; Covariants, Contravariants and the Metric Tensor

It is basic knowledge, that the brain deals with external physical objects (such as movements), which are invariant to the existence or characteristics of the systems of coordinates in which they may be expressed. Tensor network theory was established during the past years, with the contribution of R. Llinàs, based on the axiom, consistent with the above notion, that within the brain these physical invariants are expressed in natural non-orthogonal frames of reference, that are intrinsic to the organism. The proposal that neuronal networks should be considered as tensors (Pellionisz and Llinàs 1979) utilized the conceptual definition that tensors are mathematical entities expressing invariants in any frame of reference in a generalized vectorial manner (cf. Coburn 1970, Levi-Civita 1926, Wrede 1972, for other coordinate-free representations see list in Pellionisz 1983, or Lefschetz 1977).

The basic concepts and formalisms of tensor network theory emerged in a series of papers by Pellionisz and Llinàs (1979, 1980, 1982, 1985). A fundamental mathematical distinction was that in non-orthogonal intrinsic frames there are two kinds of expressions, covariant and contravariant vectors, which can be assigned to the same object (cf. Einstein 1916, Levi-Civita 1926, Coburn 1970, Wrede 1972). This notion was introduced into neuroscience in Pellionisz and Llinàs (1980; see Fig. 3 in that reference). It was pointed out that covariant expressions, composed of projection-components, are characteristic to sensory measurements in the CNS, while for motor execution parallelogram-components (contravariants) are required (see Fig. 3). These two versions can be transformed from one another by a metric tensor, implemented by a neuronal network (cf. Fig. 4). As shown in Fig. 4, the neuronal implementation of tensor expressions can be presented at multiple levels of abstraction; in abstract tensor notation, particular numerical matrix expression, network representation, graphic visualization and verbal description.

2.3 Intrinsic Vectorial Representation of Invariants Pose Problems That Can Be Resolved by Tensor Theory

Once the notion is accepted that vectorial transformations within and among intrinsic frame of reference could be adequately described by coordinate-system-general tensor analysis, several problems of such vectorial representations can be resolved as follows.

2.3.1 Intrinsic Coordinate Systems Are Non-Orthogonal

Systems of coordinates intrinsic to CNS are usually not rectangular Cartesian frames (see Figs. 2 and 3). Therefore, a method such as tensor analysis, which describes transformation of vectors expressed in different general (non-orthogonal) coordinate frames, is a natural choice (Pellionisz and Llinàs 1980).

2.3.2 CNS Uses General (e.g. Space-Time) Coordinates

The general character of intrinsic coordinates is underscored by the very profound notion and findings (cf. Braitenberg 1983, Behrend 1984, Baker et al. 1985) that information on space and time is jointly handled within the brain. This principle is evidenced e.g. by the fact that secondary vestibular neurons in the VCR have been found to express not only space, but spacetime information (Baker et al. 1985). As elaborated in Pellionisz and Llinàs 1982, tensor theory as means of treating general systems of coordinates is capable of handling transformations including spacetime representations.

2.3.3 Systems of Coordinates Are Overcomplete

The number of axes in motor systems of coordinates is often higher than the number of axes in the sensory frame (and often higher than the degrees of freedom): see the transformation to six rotational axes of extraocular muscles from three vestibular canals in Fig. 2, or the overcomplete transformation from 2-to-3 axes in Fig. 3. As a result, the transformation in such a system from sensory input to motor output is not unique. When facing any quantitative model of neuronal processing, this problem of overcompleteness is a major hurdle beyond the non-orthogonality of intrinsic frames of reference. For a full elaboration of a tensorial solution of this general problem of coordination (i.e. how the CNS may arrive at a unique choice) see Pellionisz (1984).

The explanation is briefly as follows (see Fig. 3). Similarly as shown by an example in Pellionisz and Llinàs 1980, 1982, a displacement in the two-dimensional plane can be expressed in a complete frame with two axes (left two diagrams in Fig. 3), and in an overcomplete frame with more than the necessary number of axes, e.g. three-axes (see right two diagrams in Fig. 3). The proposed three-step procedure (elaborated in Pellionisz 1984) is a covariant-contravariant transformation in the sensory frame, a sensorimotor transformation by covariant embedding, and a covariant-contravariant transformation in the motor frame. This scheme makes quantitative, experimentally testable predictions of signals at all stages of the process. As is shown later in this paper, the general scheme of Fig. 3 is the basis upon which particular elaborations of specific sensorimotor networks can be made, given that quantitative measurements of their intrinsic coordinate systems are available.

2.3.4 Subsystems That Used to Be Regarded as Separable Are Interconnected (e.g. by Convergences)

Although Sherrington (1906) emphasized the interconnected character of "compound reflexes", for practical reasons subsystems in the CNS have usually been regarded as "separable reflexes", such as the horizontal, vertical and torsional VOR. However, data demonstrate a high degree of convergence from several vestibular canals to single secondary vestibular neurons (Curthoys and

Markham 1971, Markham and Curthoys 1972). Such interconnections are not only theoretically required by tensor network theory but the necessary convergences are quantitatively predicted (see Fig. 4).

2.3.5 Geometrical Axes and Optimal Activation Axes of Effectors Are Different

A surprising experimental finding is (cf. Baker et al. 1985) that a muscle often exhibits maximal EMG activity in response to a physical rotation along a different axis than the rotation-axis belonging to the muscle-pull. This difference follows from and can be quantitatively predicted by tensor theory (Peterson et al. 1985). As quantitatively elaborated in Pellionisz et al. (1986), the predicted and measured deviations provide a reasonable match in the case of vestibulocollic reflex (VCR).

2.4 Two Basic Pillars of Brain Theory: Explaining Sensorimotor Operations, in Terms of Neuronal Networks

The tensorial approach to brain function is primarily concerned with sensorimotor networks.

The reason for concentrating on *sensorimotor* systems is that the suitability of the tensorial approach both to provide a solid basis for brain theory and to resolve problems in vectorial representations in general intrinsic frames of reference, can be best demonstrated in these basic CNS paradigms. More specific arguments are as follows.

1. Since the primary purpose of the CNS is to adequately interact with the physical world, brain function can be most readily characterized by the manner of how the CNS senses the physical environment and how it responds to it by motor action. Thus, any understanding of CNS function has to explain first such basic brain functions such as sensorimotor operations.
2. Sensorimotor functions are physically evident. They can be experimentally understood in purpose, structure, and in basic paradigms by physical means. Thus, through the interpretation of sensorimotor operations, brain theory can be best established as "neurophysics", an integer part of the exact mathematical branch of biology: that is, biophysics.
3. Brain theory is best anchored to experimental neurobiology if it provides a sensorimotor theory. Several theories of brain function are concerned with higher mental functions without, however, providing direct insights to such basic CNS function as sensorimotor operations. Before answering the question of how higher mental functions are conducted inside the head, systematic brain theory needs to explain first how the head itself can be held steady in a moving world. Primarily, brain theory ought to explain, in a clear physical manner, these relatively simple problems

e.g. how basic sensorimotor systems – such as the vestibulo-ocular and vestibulo-collic apparatus – can stabilize gaze by eye and/or head movements. Following such sequence might best connect brain theory to the neurosciences and then theory might be understood the easiest possible manner.

The reason why the tensorial approach focuses on explanation of brain function in terms of *neuronal networks*, is equally strong. Today, it is a trivial statement that the "atoms" of CNS function are the myriads of "single neurons" (although today's taken-for-granted truism had been yesterday's acrimonious controversy, cf. Gogi versus Cajal; see in Santini 1975). Presently, one of the most basic problems in neuroscience is how to interrelate the activities of "particles" of the system (the neurons) to the "mass-action" that emanates from the whole (the networks). This role of connecting single neuron and assembly-levels must be fulfilled by network theories. They should account for both the basic properties of single neurons and the function and structure of neuronal networks.

In earlier approaches, such as the "neuronic equations" (see their tensorial reformulation in this volume), this problem of connecting different representations was circumvented e.g. by stating that "our machine does not purport to realize necessarily an anatomical model of the brain, that is, there need be no one-to-one correspondence between the anatomical neuron and the basic unit of the machine" (Caianiello 1961). Beyond the problem of not adhering to the structuro-functional identity of single neurons, there is a major difficulty inherent in the fundamental assumption of the class of neuronal modeling spearheaded by the above manifesto. The whole McCulloch-Pitts school (McCulloch and Pitts 1943) is built on the axiom that the neurons are binary (1 or 0) elements, thus the natural choice is the formalism of Boolean algebra (developed for binary elements of digital computers). A tensorial reinterpretation of this major school of thought runs counter to that fundamental assumption, since if a functional space of a population of n neurons is represented by an n-space with discrete mathematical points (representing all or none activities over each neuron), then such a "non-smooth" discrete manifold is not derivable.

It is interesting to note that the other major school of neuron-population modeling, the alternative approach to McCulloch-Pitts theories (McCulloch and Pitts, 1943), is trademarked by another paper of the same pioneering team (Pitts and McCulloch 1947). This approach is based on the application of vectorial formalism. Vectors, however, may either be applied to externally characterize some overall functional properties of entire systems or subsystems, without breaking the whole network into its "particles" (somewhat similarly to reticular theories), or may represent activity-levels (e.g. firing frequencies) of a set of n neurons by a mathematical point in an n-dimensional manifold (see Kohonen 1972, Amari 1977). In either case, however, it is usually assumed that the frames of reference are extrinsic (e.g. orthogonal Cartesian) frames, where distinction of covariants and contravariants is superfluous. Indeed, in such approaches neither the frames of reference, nor the invariants that such mathematical vectors represent are explicitly identified. A tensorial reinterpretation of such approach is possible, however, (since the manifold is reasonably smooth) by identifying both the physical invariants, and the generally non-orthogonal intrinsic frames of reference in which they are expressed. This would necessitate, however, the further distinction of covariant, contravariant (and mixed) vectorial expressions.

It may be pointed out that the tensorial approach unifies the abstract vectorial representations with the concrete physical invariants of the world (that are external to CNS) through the instruments of intrinsic frames of reference. It also connects abstract vectors (activities of neuronal elements) to the anatomical reality of neuronal networks, which implement transformations on such vectors. Such networks often form anatomical "modules", cf. the lookahead-module in Pellionisz and Llinàs (1979), or the covariant-embedding-modules in Pellionisz (1984). Therefore, the tensor approach, by representing activity vectors as tensors of rank one and networks as tensors of rank two, provides means of unification of abstract vectorial approach, used in physiology, with the concrete "modular" approach, utilized in morphology.

2.5 Quantitative Elaboration of the Basic Tensorial Sensorimotor Scheme

2.5.1 Covariant-Contravariant Transformation in Overcomplete Sensorimotor Systems by Means of the Moore-Penrose Generalized Inverse

When elaborating the general tensorial sensorimotor scheme shown in Fig. 3 into a quantitative sensorimotor model, in case of overcompleteness the crucial step is to determine the covariant-contravariant motor metric transformation. As discussed in Pellionisz (1984), the problem of coordination in overcomplete systems may be resolved by means of the concept of generalized inverses (Albert 1972, Kohonen 1972, Ben-Israel and Greville 1980). Mathematically speaking, the problem is not that the inverse of the covariant metric does not exist in overcomplete systems, but that there are an infinite number of inverses. Biologically, the problem is to determine on what principles the CNS arrives at its single choice of inverse when executing a movement. Tensor network theory focuses on the hypothesis that one of the special mathematical solutions, the Moore-Penrose generalized inverse may be selected by the CNS from the infinite possible choices.

Once this hypothesis is assumed, the tensorial scheme can be elaborated into an abstract, yet quantitatively stated, general tensor network model of sensorimotor operations (Pellionisz 1984). Based on this general solution, particular applications can be numerically elaborated, as shown in the rest of the paper. These examples are based on studies on the VOR in human (Pellionisz 1985b), VOR in rabbit (Simpson and Pellionisz 1984), VOR in cat (Pellionisz and Graf 1986), VCR in human (Pellionisz 1985a), and VCR in the cat (Pellionisz et al. 1985).

As has been explicitly stated (e.g. in Simpson and Pellionisz 1984), there may be other solutions utilized by the CNS that employ special generalized inverses other than the one mathematically established by Moore-Penrose. However, tensor network theory offers some rather substantial arguments indicating that the Moore-Penrose hypothesis should be tested first. First, the Moore-Penrose inverse has numerous mathematically attractive features. For instance, it is a least-squares optimum formula so that it implies efficient use of energy by minimizing the overall activation of muscles acting about a joint. Second, and perhaps more important, the neuronal network that may implement this special kind of transformation in the CNS can be constructed in the nervous system in a straightforward manner. As discussed later in this paper and elaborated in detail (Pellionisz and Llinàs 1985), the proposed "Metaorganization" method (a) is based on a physically simple procedure of oscillations, (b) is biologically plausible requiring only reverberations of proprioception – motor execution (c) requires no "computational depth" in neuronal networks (arriving at a stabilized oscillation and invoking proportional connections in an array of neurons, e.g. in the cerebellum).

The assumption that the CNS utilizes a solution in accordance with the above tensorial suggestions of intrinsic coordinates and Moore-Penrose generalized inverse has been substantiated recently by Gielen and Zuylen (1986).

3 Particular Examples of Sensorimotor Modeling: Tensor Network Models of the Gaze Stabilization Mechanisms

Sensorimotor systems offer special advantages for demonstrating the quantitative applicability of brain theories, since the input and output of such mechanisms are physically well known, the overall function of the transformation is evident, and the basic structural features of the neuronal networks which connect the input to the output are usually also rather well known. Thus, once the frames of reference intrinsic to the transformation from sensory to motor expression are available, the general three-step scheme of sensorimotor transfer (shown in Fig. 3) can be quantitatively elaborated. To demonstrate such applicability of the general concepts, and to show that the resulting models can lead to experimentally testable paradigms, some worked-out examples are shown below.

3.1 Gaze Stabilization by Vestibulo-Ocular Reflex: A Tensor Network Model of the "Three-Neuron Reflex-Arc" in the Cat

The model offered in Pellionisz (1985b) was the first quantitative elaboration of the above general concepts and formalisms to an existing sensorimotor system. The scheme was applied to calculate numerically the matrix-components of a three-step tensorial scheme of the human vestibulo-oculomotor reflex, which uses intrinsic frames of reference that had long been quantitatively established (Helmholtz 1896, Blanks et al. 1975). The model presented (in Fig. 7 in Pellionisz 1985b) demonstrated that an entire sensorimotor system can be represented once the intrinsic frames of reference are provided by experimental investigation. With similar sets of data emerging from recent experimental measurements (Simpson et al. 1981, Simpson 1983, Ezure and Graf 1984), this quantitative technique has also been applied to the VOR of other species, such as the rabbit (Simpson and Pellionisz 1984) and the cat (Pellionisz and Graf 1986).

In this paper, a tensor network model of the three-neuron vestibulo-ocular reflex-arc of the cat is shown to represent this class of sensorimotor modeling (see Fig. 4). The scheme shown in Fig. 4 (Pellionisz and Graf 1986) provides a quantitatively realistic representation of the neuronal networks necessary for transforming the vestibular semicircular canal coordinates (from Blanks et al. 1972) into extraocular muscle coordinates (from Ezure and Graf 1984). Both these sensory and motor systems are similar to the one shown, for the human, in Fig. 2. The model closely follows the three-step VOR scheme for human (Pellionisz 1985b). However, aiming at a description that more closely approximates the actual morphological implementation of the vestibulo-ocular reflex, in the model of the VOR in cat the first two transformations (the sensory metric and the sensorimotor covariant embedding) were contracted. This corresponds to the fact that direct connections exist from secondary vestibular nuclei to oculomotor neurons (Lorente de Nó 1933, Szentàgothai 1950). Thus, the sensory metric and covariant sensorimotor embedding transformations may be implemented in the CNS by a single step, yielding a "three-neuron reflex-arc". The connectivity predicted by the model is shown in Fig. 4 in general tensor notation, in particular matrix expression, in corresponding dot-diagrams, and in quantitatively realistic (thickness-coded) network diagram. Such network-models can be directly compared with detailed morphological features of the vestibulo-ocular projections revealed by intracellular injection of horseradish peroxidase (McCrea et al. 1981, Ezure and Graf 1984). The close correspondance, as discussed in Pellionisz and Graf (1986) indicates that the model realistically captures the important features of the 3-neuron vestibulo-ocular reflex arc.

Fig. 4. Tensor network theory yields a quantitative and morphologically realistic model of the basic sensorimotor apparatus, the "three-neuron vestibulo-ocular reflex-arc" (after Pellionisz and Graf 1986). The general scheme of Fig. 3 is elaborated here for a sensorimotor system with known frames of reference. The three rotational-axes of the vestibular sensory – and six axes of extraocular motor apparatus are quantitatively available; see axes for vestibular semicircular canals (*A* anterior; *H* horizontal; *P* posterior), and for extraocular muscles (*LR* lateral rectus; *MR* medial rectus; *SR* superior rectus; *IR* inferior rectus; *SO* superior oblique; *IO* inferior oblique). Tensor-transformation matrices, corresponding to Fig. 2, have been calculated for these frames. The sensory metric and sensorimotor embedding tensors are contracted, thus a "three-neuron-arc" model emerges; a vestibular tensor transformation v^r_i and an oculomotor tensor transformation g^{ie}. Transformations are also shown in the form of matrices, dot-diagrams, and a quantitatively precise network-diagram, a realistic representation of the existing neuronal mechanism in the CNS (Pellionisz and Graf 1986)

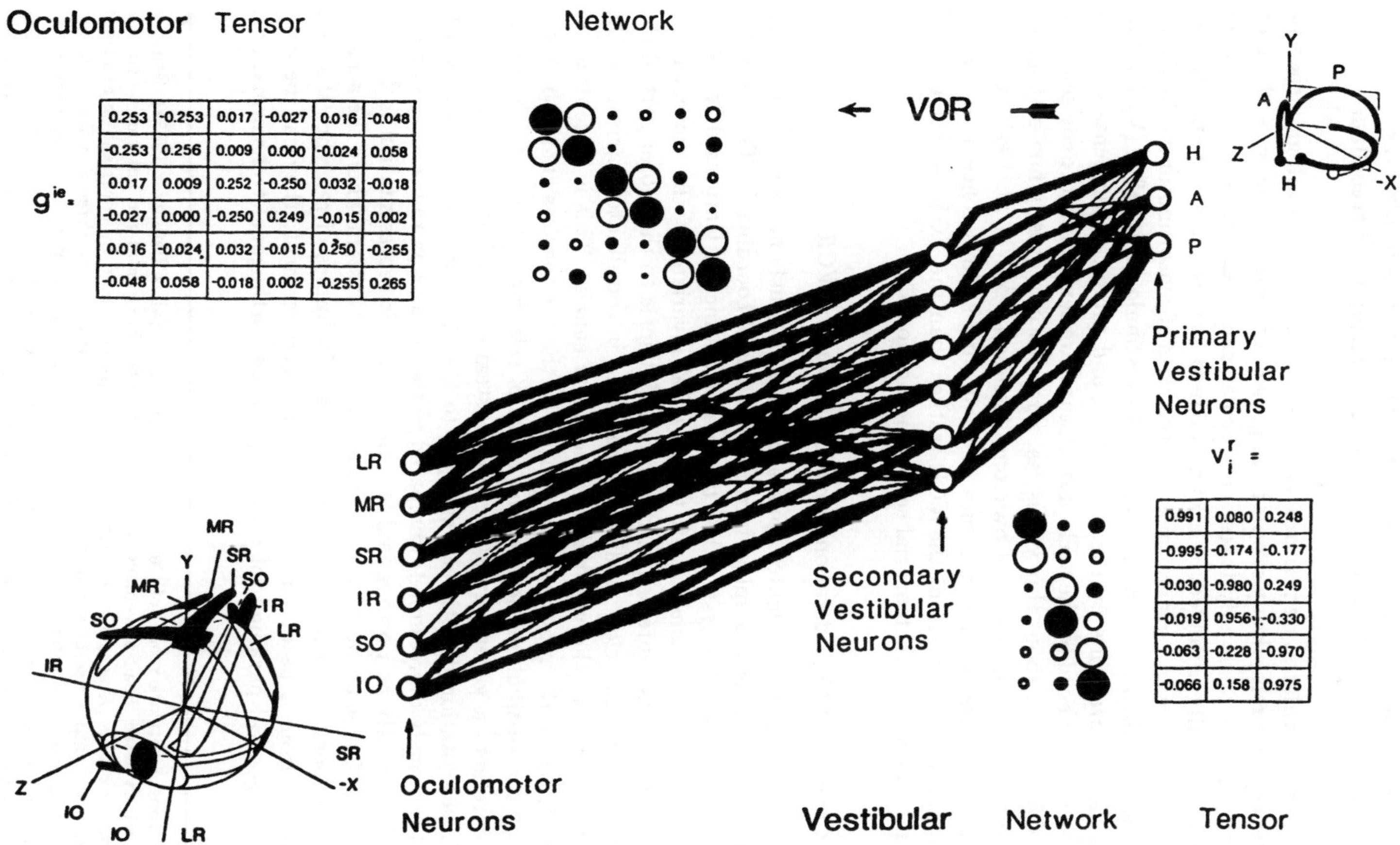

$g^{ie} =$

0.253	-0.253	0.017	-0.027	0.016	-0.048
-0.253	0.256	0.009	0.000	-0.024	0.058
0.017	0.009	0.252	-0.250	0.032	-0.018
-0.027	0.000	-0.250	0.249	-0.015	0.002
0.016	-0.024	0.032	-0.015	0.250	-0.255
-0.048	0.058	-0.018	0.002	-0.255	0.265

$v_i^r =$

0.991	0.080	0.248
-0.995	-0.174	-0.177
-0.030	-0.980	0.249
-0.019	0.956	-0.330
-0.063	-0.228	-0.970
-0.066	0.158	0.975

Fig. 4

3.2 Tensor Network Model of the Head-Stabilization Sensorimotor Mechanism of the Vestibulo-Collic System, Involving the Cerebellum

The vestibulo-ocular system is probably the best example of a highly specialized system with remarkably simple arrangement of both the sensory and motor elements. Indeed, with only three semicircular canals and six eye muscles, all quasi-orthogonally arranged in a quasi-aligned manner, the system may be too simple to serve as a general model of the class of sensorimotor mechanisms. Other gaze-stabilization systems, such as the vestibulo-collic reflex (VCR) that detects head-movements by the vestibulum, and produces compensatory displacement of the head by the neck-musculature, are certainly more complicated. The VCR involves about thirty muscles, arranged in a grossly non-orthogonal manner, with a high degree of overcompleteness (Baker et al. 1985). Beyond this considerations on the structure, the following serious conceptual reasoning supports the selection of the VCR (rather than the VOR) as the optimal candidate for a testable tensorial sensorimotor model.

3.2.1 Conceptual Considerations for Focusing on the VCR

It is a closed-loop sensorimotor apparatus with obviously non-orthogonal, multidimensional and overcomplete intrinsic systems of coordinates. Theoretically, the VCR sensorimotor system has two major distinctive features differentiating it from other systems. First, the physical invariant of the movement of the head, (which is expressed in coordinates intrinsic to the motor part of the system, the neck muscles), is directly measured in coordinates intrinsic to the sensory apparatus, (the vestibular semicircular canals). This is in contrast to the VOR, where the movements produced by the eye muscles cannot be measured by the vestibulum. Second, the frame of reference intrinsic to the motor part represents a general case of motor system of coordinates, it consists of axes (head-rotation directions, belonging to the 30 individual muscles) which are (a) obviously overcomplete compared to the six axes in the sensory system, and (b) obviously non-orthogonal (see the schematic representation in Fig. 5). Thus, the VCR is an optimal primary sensorimotor system where the sensory and motor portions express the same invariant in frames of reference that are far enough from a pseudoorthogonal, paired arrangement; they can be considered a general paradigm of overcomplete non-orthogonal transformations.

Fig. 5. Tensor network model of the vestibulo-collic reflex (after Pellionisz 1985a). Sensory signals which detect head-movements are provided by the horizontal and anterior vestibular semicircular canals. A sensorimotor transformation (implemented by covariant embedding) converts them into a vectorial expression of the movement in the three-axis motor frame (representing the head-rotation axes as a result of the pulling of three exemplary neck-muscles). Sensorimotor embedding, however, yields covariant (orthogonal projection type) vectorial expression, whose execution will not yield the required rotation. The covariant intention – contravariant execution transformation is provided by the cerebellar neuronal network (for detailed explanation, see Pellionisz 1984, 1985a)

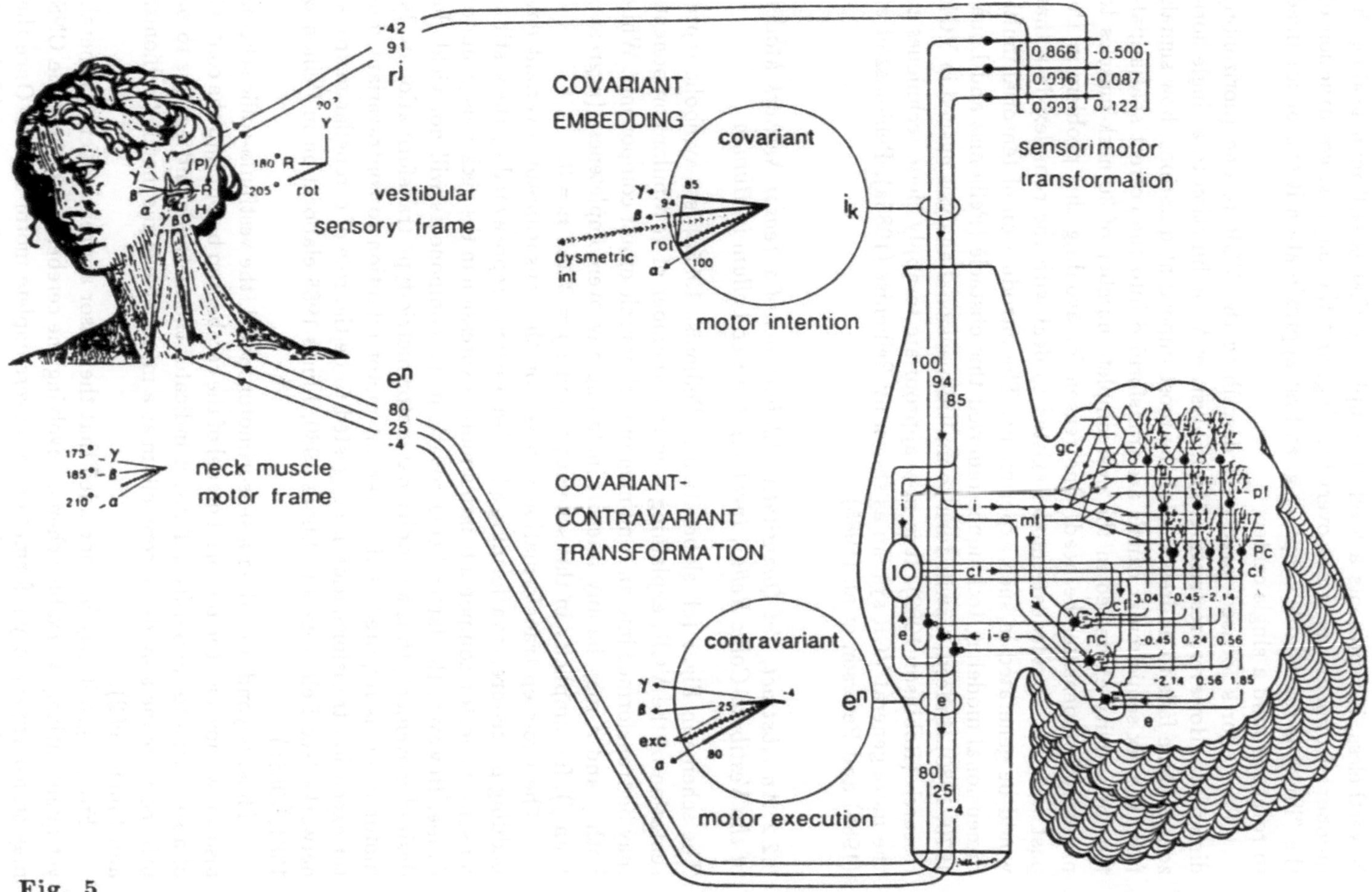
-42
91
r^j
vestibular
sensory frame
e^n
80
25
-4
neck muscle
motor frame
COVARIANT
EMBEDDING
covariant
dysmetric
int
i_k
motor intention
COVARIANT-
CONTRAVARIANT
TRANSFORMATION
contravariant
e^n
motor execution
0.866 -0.500
0.996 -0.087
0.993 0.122
sensorimotor
transformation
IO
gc
mf
pf
Pc
cf
nc
i-e
3.04 -0.45 -2.14
-0.45 0.24 0.56
-2.14 0.56 1.85

Fig. 5

Nevertheless, it represents a level of complexity that is still manageable; the number of dimensions is not overwhelming, and the physical construction of the "limb" is reasonable simple, e.g. at first approximation it can be assumed to rotate around a single center.

Past efforts focused largely on describing the VCR in one (horizontal) dimension. However, restricting the analysis of VCR function to a single horizontal plane fails to address one of the most important questions: how signals from the six semicircular canals are transformed into the correct spatial pattern of activity expressed in the overcomplete number of 30 neck-muscles to precisely compensate the head. One reason for avoiding this problem in the past was that no adequate model existed to deal with the complex data that would be generated in such experiments. The introduction of tensorial multidimensional modeling techniques removed this obstacle (Pellionisz and Llinàs 1979, 1980). Indeed, a swift turn to multidimensional analysis ensued in VOR research (Robinson 1982). It is now appropriate to apply these techniques to the more general VCR system as shown by Pellionisz (1985a), Pellionisz et al. (1986) and Peterson et al. (1985).

3.2.2 An Abstract, Yet Quantitative Elaboration of a Tensor Network Model of the Vestibulo-Collic Reflex, Involving the Cerebellum in Human

The scheme in Fig. 5 (cf. elaboration in Pellionisz 1985a) is a symbolic representation of the VCR, explaining how the j-number of vestibular components may be transformed into an n-number of neck-muscle motor components. While both j and n may be any integer, including any overcompleteness (n greater than j), for simplicity in this schematic model $j = 2$ and $n = 3$.

The two-step transformation is based on the sensorimotor covariant embedding procedure, transforming a head-movement, represented by the vestibular sensory vector-components into a motor expression in the neck-muscle motor frame. However, the latter being covariant, its components will not yield the desired movement, thus an additional motor-metric-type transformation in the motor frame is required. This covariant motor intention to contravariant motor execution transformation is suggested to be the role of cerebellar neuronal networks (see Pellionisz and Llinàs 1980, Llinàs 1985 elaboration in Pellionisz 1984, 1985a).

Thus, beyond providing a sensorimotor model, the vestibulo-collic scheme also shows how can the tensorial model of the cerebellum be put into the context of a sensorimotor apparatus. Thus, it indicates the methods and designs to be utilized to refine a model of cerebellum as a motor spacetime metric (Pellionisz and Llinàs 1982).

From Fig. 5 it may be concluded that the tensor approach offers a general, yet quantitatively workable scheme, involving the cerebellum, of how the CNS may transform sensory information into overcomplete motor signals. Once the frames of reference of the neck-muscle were quantitatively revealed (Baker et al. 1985), the transformation-matrices necessary for such conversion could be calculated, and their experimental tests are presently in progress (cf. Peterson et al. 1985, Pellionisz et al. 1986). Agreement between initial predictions of spatial properties of neck motor signals and experimental measurements is close enough to suggest that a model based on the design described above will be able to quantitatively account for the transformations underlying the cat's VCR (Peterson et al. 1985). Such model is obviously much more complex than that of the VOR, yet the computation is feasible by the use of microcomputer modeling techniques.

While the foregoing presentation has concentrated upon the properties of the VOR and VCR, the general applicability of the approach is obvious. Most sensorimotor systems in the body involve sensory receptors with a wide variety of spatial orientations providing signals to control the action of complex multi-arcticular structures such as neck or limbs whose movement is determined by a collection of muscles whose number exceeds the number of degrees of freedom controlled by the system. To understand these systems, we need experiments based upon multidimensional models capable of incorporating the actual structural complexity, such as overcompleteness of the control system. These types of experiments, concerned with movements of the forelimb (Soechting and Ross 1984, Kolb and Rubia 1985, Gielen and Zuylen 1986), and those conducted for VCR (Baker et al. 1985, Peterson et al. 1985) based on tensorial modeling, represent the first definitive tests of a new model-system designed to meet this challenge. If basic properties of systems such as VOR and VCR can be analyzed and understood with these techniques, the way will be open to applying them more broadly to solve some of the most important problems facing neuroscientists interested in motor control. A successful application of a brain theory for sensorimotor systems may be considered an essential test of its viability for further applications explaining higher mental functions.

4 Interaction of Physical and Functional Geometries in Sensorimotor Operations and in Higher Brain Function: The Principle of Metaorganization

Sensorimotor modeling demonstrates the capability of tensor theory to explain how neuronal networks, once they are available, can perform parallel transformations yielding the necessary overall function. However, it is immediately obvious that even a successful modeling of the function itself would have to be followed by further studies of (a) how such sensorimotor neuronal networks are generated (and modified) in the CNS, (b) how can the CNS utilize such processes of genesis and modification in mental functions that are much beyond the primary sensorimotor level. These questions have recently been addressed by the development of tensor network theory; by proposing the principle and procedure of Metaorganization (cf. Pellionisz 1983, Pellionisz and Llinàs 1985). Below, some of the ideas covered there, and some that carry those studies further, are briefly exposed.

4.1 Genesis and Modification of the Metric of Sensorimotor Systems

Tensor network theory regards the central nervous system as a geometrical object. This idea led to representations of basic sensorimotor systems by tensor network models. Such networks implement metric tensors in the frames of references intrinsic to both the sensory and motor apparatus. Each of these motor and sensory metric tensors comprise a functional geometry, which is an internal representation of the physical geometry of the motor apparatus, and of the perceived external physical world, respectively. From a more general viewpoint, brain function can, therefore, be interpreted as the matching of the relationships existing in the external world with their internal CNS representation. In the interaction of these sets of relationships, defined as the process of Metaorganization, physically explicit external "body" geometries and the internal CNS functional geometries mold one another. Most particularly, the structural geometry of the motor system interacts, by means of neuronal pathways, with a functional geometry implemented by the cerebellar neuronal network.

This interaction is symbolically represented in Fig. 6, where the frame of reference, intrinsic to a three-joint limb is shown by alpha, beta, gamma local-displacement axes. As discussed in Pellionisz and Llinàs (1985), the motoneuron activation vector of this system, and the corresponding muscle proprioception-vector represent contravariant and covariant expressions of the limb activation. The conceptual basis of Metaorganization is this dual vectorial expression of physical objects by covariant sensory and contravariant motor vectors. These dual vectorial expressions are rendered to one another as determined by (a) the physical geometrical arrangement of the motor system external to the CNS, (b) the neuronal network of the cerebellum which acts as a metric tensor, internal to the CNS.

As depicted in Fig. 6, the contravariant motor execution (e) generates a physical action on the external world via the musculo-skeletal system. The covariant sensory proprioception measures the effect of such motor output (p). This transformation from motor execution to proprioceptive sensory detection, expressible by a covariant metric tensor g_{nk}, is implicit in the physical geometry of the motor system. However, in order to generate a coordinated, goal-directed motor action, the other type of metric transformation, given in a contravariant form, must be made available. This contravariant metric is implemented by the cerebellar network which turns a covariant motor intention vector into a contravariant motor execution vector.

The proposed physical process of the implementation of the Metaorganization principle is based on oscillatory reverberations. Covariant proprioception (p) is used as a signal recurrent to the motor apparatus, as if it were a contravariant motor expression (e). The steady-state of such reverberations yields

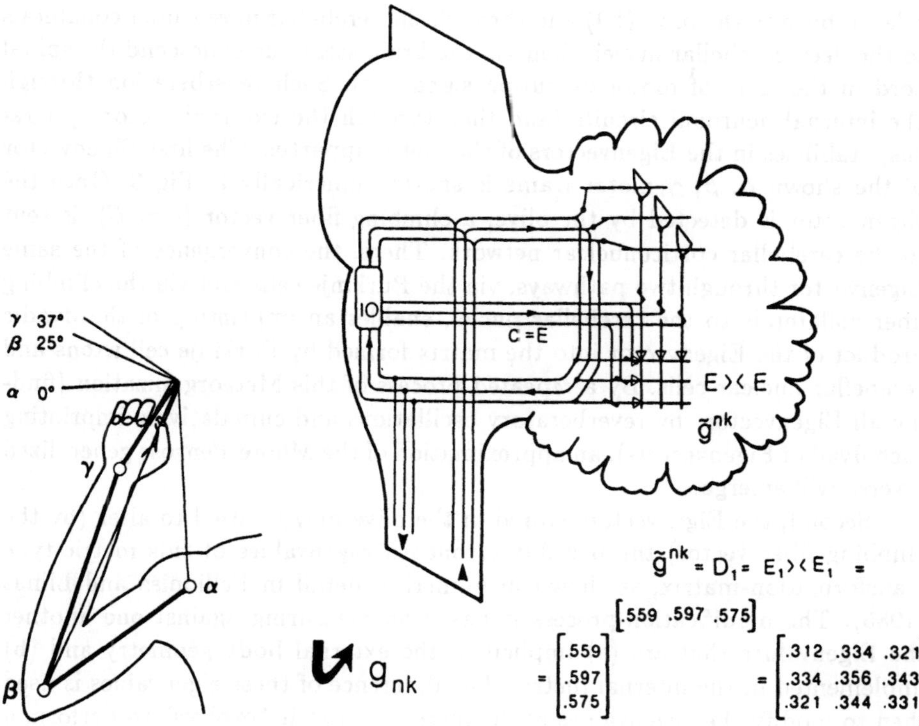

Fig. 6. Metaorganization of a functional geometry by a physical geometry (after Pellionisz and Llinàs 1985). The system of coordinates intrinsic to a simplified motor system of a three-joint limb are shown by the α, β, γ local displacement-axes (in response to joint-angle changes), forming a non-orthogonal overcomplete frame of reference. Motor signals, deflecting joint-angles, descend the spinal cord in the form of a motor vector e, which is contravariant since its components physically execute the movement. In turn, the muscle proprioception-vector p, provides a measure, a covariant representation of the movement. This vector ascends the spinal cord; the physical geometry of the executor mechanism in effect implements a covariant metric transformation from e to p by g_{nk}. Reverberation of e to p through the physical g_{nk} (via mossy fiber collaterals to cerebellar nuclei and via the brain-stem nuclei) arrives at Eigenvectors (when $e = p$ in a normalized form, see an actual Eigenvector shown numerically). Eigenvectors are detected by and stored in the inferior olive, and sent (via $c = E$ climbing fiber vector) to the matrix formed by Purkinje cell axons and the dendritic trees of cerebellar nuclear cells, imprinting the dyadic product D of the Eigenvectors. Having found and imprinted all dyads of Eigenvectors, the approximation of g^{nk}, as the Moore-Penrose generalized inverse of g_{nk} emerges. (For detailed explanation, see Pellionisz and Llinàs 1985)

the Eigenvectors (E) and Eigenvalues (λ) of the system (cf. Brogan 1974). These values were proposed to be stored in sub-cortical nuclear networks, such as the inferior olive (IO). The stored Eigenvectors and Eigenvalues then serve two purposes as follows.

First, the Eigenvectors are the means for the genesis of a metric in the form of spectral representation of the Moore-Penrose generalized inverse of the covariant metric (g^{nk}). Essential elements of this process are depicted in Fig. 6 by showing the reverberatory pathways; the ascending proprioception (p) which is relayed both to the olive (IO) and through the cerebellar mossy fiber collaterals to the deep cerebellar nuclei, then via the brain stem nuclei descend the spinal cord in the form of motor execution signals (e). Such reverberation through the internal neuronal circuits (and then through the external motor apparatus) stabilizes in the Eigenvectors of the motor appartus. The first Eigenvector of the shown α, β, γ motor frame is shown numerically in Fig. 6. Once the Eigenvector is detected by the olive, a climbing fiber vector ($c = E$), is sent to the cerebellar corticonuclear network. There, the convergence of the same Eigenvector through two pathways, via the Purkinje cells and via the climbing fiber collaterals to the cerebellar nuclei, enables an imprinting of the dyadic product of the Eigenvector into the matrix formed by Purkinje cell axons and cerebellar nuclear cells. By a repeated process of this Metaorganization (finding all Eigenvectors by reverberatory oscillations and cumulatively imprinting each dyad of Eigenvectors), an approximation of the Moore-Penrose generalized inverse will emerge.

Second, the Eigenvectors stored in the olive may be used to alter (by the climbing fiber vector) the possibly erroneous Eigenvalues of this metric-type transformation-matrix, as shown in numerical detail in Pellionisz and Llinàs (1985). The modification-process is based on measuring against one another the Eigenvalues that are (a) implicit in the external body geometry and (b) implemented in the internal metric. The difference of these eigenvalues is used then to modify the Eigenvalues of the metric so that it "evolves" to perform a perfected covariant-contravariant transformation.

4.2 Metaorganization of Hierarchical Functional Geometries

The process of Metaorganization was shown above to generate and modify a neuronal network comprising the functional counterpart of a physical geometry. However, as shown in Pellionisz and Llinàs (1985), the principle leads to far-reaching generalizations. Their basis is the fact that the external system that converts a contravariant signal (such as e) to its covariant counterpart (p) need not be a physical apparatus, but can also be another *functional* geometry. This can be comprised by another neuronal network, or by a combination of a physical geometry and a functional geometry, implemented by a network.

According to the above, once the motor metric of the cerebellum is generated by Metaorganization, and the sensorimotor covariant embedding transformation is also implemented by a simple procedure, described by Pellionisz (1984) and Pellionisz and Llinàs (1985), there is a possibility of utilizing for reverberation the compounded entire sensorimotor circuit (see Fig. 7). This entire sensorimotor circuit consists of the covariant sensorimotor embedding, the motor metric, the physical geometry of the motor system and the physical geometry of the sensory system. The reverberation though this whole circuit can Metaorganize a sensory metric of the entire system. This can be implemented (as shown by the scheme in Fig. 7), by issuing s^j contravariant sensory vectors, which would be transformed (via c_{jk} embedding, g^{nk} motor metric, g_{nk} motor

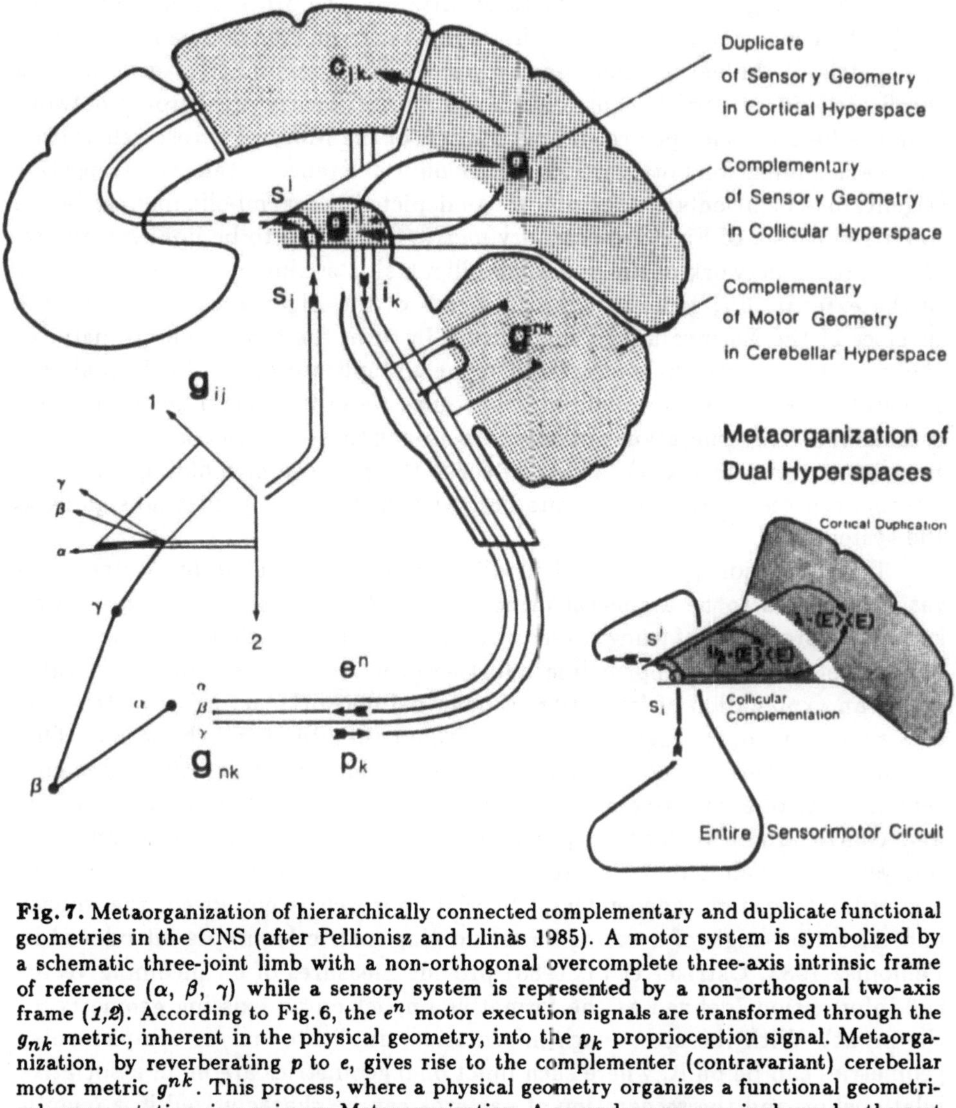

Fig. 7. Metaorganization of hierarchically connected complementary and duplicate functional geometries in the CNS (after Pellionisz and Llinàs 1985). A motor system is symbolized by a schematic three-joint limb with a non-orthogonal overcomplete three-axis intrinsic frame of reference (α, β, γ) while a sensory system is represented by a non-orthogonal two-axis frame (*1,2*). According to Fig. 6, the e^n motor execution signals are transformed through the g_{nk} metric, inherent in the physical geometry, into the p_k proprioception signal. Metaorganization, by reverberating p to e, gives rise to the complementer (contravariant) cerebellar motor metric g^{nk}. This process, where a physical geometry organizes a functional geometrical representation, is a primary Metaorganization. A secondary process is shown by the rest of the Figure (see also *inset* in *lower right*), where the entire sensorimotor circuit acts as the g_{ij} metric that transforms every s^j contravaraint sensory vector into its covariant s_j counterpart. Reverberation through the entire circuit yields the Eigenvectors (*E*) and Eigenvalues (λ). Dyadic products of these Eigenvectors, with λ and λ^{-1} coefficients, give rise to either or both a complementer and a duplicate metric-type geometrical representation. (For detailed explanation, see Pellionisz and Llinàs 1985)

geometry, and g_{ij} sensory geometry) into a covariant sensory reception vector s_i. Thus, through the entire system, to every contravariant s^j, its covariant s_j can be obtained. According to the Metaorganization-principle, such rendering via oscillatory reverberation is sufficient to establish the Eigenvectors of the entire sensorimotor circuit. Once the Eigenvectors are established, according to an interesting feature of the Metaorganization, the process yields an opportunity to construct either (or both) a covariant- or contravariant version of the metric of the entire sensorimotor circuit. This possibility of generating a duplicate and/or complementer geometry, in the form of a neuronal network emerges because the spectral representation of the Moore-Penrose generalized inverse is similar for both metrics. Their only difference is that the respective Eigenvalue-coefficients are reciprocal. As depicted in a symbolic manner in the lower right inset of Fig. 7, the sensory metric (suggested to be implemented by the neuronal network of the superior colliculus) is a complementary geometry of the external g_{ij}, generated by the dyads of the Eigenvectors by using the inverses of the Eigenvalues as coefficients. In turn, the functional geometry of the cortical hyperspace is suggested to be a duplicate of g_{ij}. Such dual representation leads to the possibility that once these functional geometries are established, a reverberation can be set up among these two metrics (symbolized by heavy arrows in Fig. 7). This can serve the purpose of molding these two internal representations to one another, without the use of external signals to the system.

The formation (genesis and modification) of the metric by Metaorganization, appears to be a general principle of CNS function. Metaorganization suggests a procedure of using reverberations for such formation. However, based on the Metaorganization principle, a further generalization is in progress, which yields an explanation of formation of neuronal networks acting as metric tensors as a result of an ongoing process, without the use of reverberations. This avenue is presently explored by means of tensorially reinterpreting some current theories of associative memory, where the explicit distinction of covariant and contravariant vectorial expressions (in non-orthogonal coordinates) has not yet been exploited. When such general frames of reference are used, and a coordinate-system-independent tensorial approach is employed, the impact and refraction-angle of a vector on an existing metric (concepts introduced in Pellionisz 1983, Ostriker et al. 1985) provide measures of an incoming vectorial information (defined as the formative impact on an existing geometry, of the input). With the use of such information, duplicate and/or complementer geometries can be built, where such functional geometries are in an associative relationship, and the recognition of patterns can be characterized in a quantitatively precise manner.

Metaorganization of metric tensors, therefore, may lead to an explanation of the emergence of hierarchically connected hyperspaces, based primarily on sensorimotor operations. Within and among them, functional geometries constantly mold one another in order to perfect the internal representations of systems of relationship external to the CNS. As a result, brain function enables the living organism to ceaselessly improve its harmony with the world. If brain theory can lead to a successful formal and quantitative elaboration of such concepts, not only our understanding of brains could be elevated to a higher level, but utilizations may result in the feasibility of constructing truly brain-like robotic systems, indeed, geometrical engines.

References

Albert A (1972) Regression and the Moore-Penrose pseudoinverse. Academic Press, London New York

Amari S (1977) Neural theory of association and concept-formation. Biol Cybern 26:175–185

Anderson JA (1983) Cognitive and psychological computation with neural models. IEEE SMC-13 (5):799–815

Baker R, Evinger C, McCrea RA (1981) Some thoughts about the three neurons in the vestibulo-ocular reflex. Ann NY Acad Sci 374:171–188

Baker J, Goldberg J, Peterson B (1985) Spatial and temporal response properties of the vestibulocollic reflex in decerebrate cats. J Neurophysiol 54:735–756

Bantli H (1972) Multi-electrode analysis of field potentials in the turtle cerebellum: an electrophysiological method for monitoring continuous spatial parameters. Brain Res 44:676–679

Behrend K (1984) Cerebellar influence on the time structure of movement in the electric fish Eigenmannia. Neuroscience 13(1):171–178

Ben-Israel A, Greville TNE (1980) Generalized inverses: Theory and applications. Krieger Publ, New York

Bernstein NA (1947) O postroyenii dvizheniy (On the construction of movements). Medgiz, Moscow

Beurle RL (1962) Functional organization in random networks. In: Foerster H, Zopf GW (eds) Principles of self-organization. Pergamon Press, New York, pp 291

Blanks RHI, Curthoys IS, Markham CH (1972) Planar relationships of the semicircular canals in the cat. Am J Physiol 223:55–62

Blanks RHI, Curthoys IS, Markham CH (1975) Planar relationships of the semicircular canals in man. Acta Otolaryngol 80:185–196

Bower J, Llinàs R (1983) Simultaneous sampling of the responses of multiple, closely adjacent, Purkinje cells responding to climbing fiber activation. Soc Neurosci Abstr 9(1):607

Braitenberg V (1983) The cerebellum revisited. J. Theor Neurobiol 2(3):237–242

Brogan WL (1974) Modern control theory. Quantum, New York

Caianiello E (1961) Outline of a theory of thought-processes and thinking machines. J Theoret Biol 2:204–235

Churchland PS (1986) Neurophilosophy: Towards a unified understanding of the mind-brain. MIT/Bradford Press

Coburn N (1970) Vector and tensor analysis. Dover, New York

Curthoys IS, Markham CH (1971) Convergence of labyrithine influences on units in the vestibular nuclei of the cat. I. Natural stimulation. Brain Res 35:469–490

Eccles JC (1973) The cerebellum as a computer: Patterns in space and time. J Physiol (London) 229:1–32

Eccles JC (1982) The modular operation of the cerebral neocortex considered as the material basis of mental events. Neuroscience 6:1839–1856

Einstein A (orig. 1916) The foundation of the general theory of relativity. In: Sommerfeld A (ed) The principle of relativity (1952). Dover, New York, pp 111–164

Epping W, Boogaard H, Aertsen A, Eggermont J, Johannesma P (1984) The neurochrome. An identity preserving representation of activity patterns from neural populations. Biol Cybern 50:235–240

Ezure K, Graf W (1984) A quantitative analysis of the spatial organization of the vestibulo-ocular reflexes in lateral and front-eyed animals. II. Neuronal network underlying vestibulo-oculomotor coordination. Neuroscience 12:95–109

Freeman WJ (1975) Mass action in the nervous system. Academic Press, London New York

Gernstein GL, Bloom MJK, Espinosa IE, Evanczuk S, Turner MR (1983) Design of a laboratory for multineuron studies. IEEE SMC 13:668–676

Gielen CCAM, Zuylen EJ (1986) Coordination of arm muscles during flexion and supination: Application of the tensor analysis approach. Neuroscience (in press)

Grimson WEL (1981) From images to surfaces. MIT Press, Cambridge

Hebb DO (1949) The organization of behaviour. Wiley, New York

Helmholtz H (1896) Handbuch der Physiologischen Optik. Voss, Leipzig

Hinton GE, Anderson JA (eds) (1981) Parallel models of associative memory. Lawrence Erlbaum Assoc, Hillsdale

Hodgkin AL, Huxley AF (1952) A quantitative description of membrane current and its application to conduction and excitation in nerve. J Physiol (London) 1176:500–544

Katchalsky AK, Rowland V, Blumenthal R (1974) Dynamic patterns of brain cell assemblies. Neurosci Res Prog Bull 12:1–187

Kohonen T (1972) Correlation matrix memories. IEEE Trans Comput C-21:353–359

Kolb FP, Rubia FJ (1985) Sensory representation of movement parameters in the cerebellar cortex of the decerebrate cat. In: Bloedel JR, Dichgans J, Precht W (eds) Cerebellar functions. Springer, Berlin Heidelberg New York, pp 282–299

Lefschetz S (1977) Differential equations: Geometric theory. Dover, New York

Levi-Civita T (1926) In: Persico E (ed) The absolute differential calculus (Calculus of tensors). Dover, New York

Lewin R (1982) Neuroscientists look for theories. Science 216:507

Llinàs R (1985) Functional significance of the basic cerebellar circuit in motor coordination. In: Dichgans J, Precht W (eds) Cerebellar functions. Springer, Berlin Heidelberg New York, pp 170–185

Llinàs R, Pellionisz A (1984) La mente in quanto proprietà tensoriale dei circuiti cerebrali. In: Palmarini MP (ed) Livelli di realtà. Feltrinelli, Italy

Llinàs R, Nicholson C, Johnson K (1973) Implantable monolithic wafer recording electrodes for neurophysiology. In: Phillips MI (ed) Brain unit activity during behavior. Charles C Thomas, Springfield, pp 105–111

Loeb GE (1983) Finding common ground between robotics and physiology. Trends Neurosci 5:203–204

Lorente de Nó R (1933) Vestibulo-ocular reflex arc. Arch Neurol Psychiatry 30:245–291

Mann RW (1981) Cybernetic limb prosthesis. Ann Biomed Eng 9:1–43

Markham CH, Curthoys IS (1972) Convergence of layrinthine influences on units in the vestibular nuclei of the cat. II. Electrical stimulation. Brain Res 43:383–396

Marr D (1982) Vision. Freeman, San Francisco

McCrea RA, Yoshida K, Evinger C, Berthoz A (1981) The location, axonal arborization and termination sites of eye-movement related secondary vestibular neurons demonstrated by intra-axonal HRP injection in the alert cat. In: Fuchs AF, Becker W (eds) Progress in oculomotor research. Elsevier, Amsterdam, pp 379–386

McCulloch WS, Pitts W (1943) A logical calculus of the ideas immanent in nervous activity. Bull Math Biophys 5:115–133

Mountcastle V (1979) An organizing principle for cerebral function: The unit module and distributed system. In: Schmitt FO, Worden FG (eds) The neurosciences; 4th Study Program. MIT Press, Cambridge, pp 21–42

Ostriker G, Pellionisz A, Llinàs R (1985) Tensorial computer model of gaze. I. Oculomotor activity is expressed in non-orthogonal natural coordinates. Neuroscience 14:483–500

Palm G (1982) Neural assemblies. Springer, Berlin Heidelberg New York

Pellionisz A (1983) Brain theory: connecting neurobiology to robotics. Tensor analysis: utilizing natural coordinates to describe, understand and engineer functional geometries of intelligent organisms. J Theor Neurobiol 2:185–211

Pellionisz A (1984) Coordination: a vector-matrix description of transformations of overcomplete CNS coordinates and a tensorial solution using the Moore-Penrose generalized inverse. J Theor Biol 110:353–375

Pellionisz A (1985a) Tensorial brain theory in cerebellar modeling. In: Dichgans J, Precht W (eds) Cerebellar functions. Springer, Berlin Heidelberg New York, pp 201–229

Pellionisz A (1985b) Tensorial aspects of the multidimensional approach to the vestibulo-oculomotor reflex and gaze. In: Berthoz A, Melvill-Jones J (eds) Adaptive mechanisms in gaze control. Facts and theories. Elsevier, Amsterdam, pp 281–296

Pellionisz A (1985c) Tensorial computer movie of the genesis and modification of cerebellar networks as dyadic expansions of the eigenvectors stored in the inferior olive. Soc Neurosci Abstr 10:540

Pellionisz A, Graf W (1986) Tensor network model of the "three-neuron vestibulo-ocular reflex arc" in the cat. J Neuroscience (submitted)

Pellionisz A, Llinàs R (1979) Brain modeling by tensor network theory and computer simulation. The cerebellum: distributed processor for predictive coordination. Neuroscience 4:323–348

Pellionisz A, Llinàs R (1980) Tensorial approach to the geometry of brain function. Cerebellar coordination via a metric tensor. Neuroscience 5:1761–1770

Pellionisz A, Llinàs R (1981) Genesis and modification of the geometry of CNS hyperspace. Cerebellar space-time metric tensor and "motor learning". Soc Neurosci Abstr 6:510

Pellionisz A, Llinàs R (1982) Space-time representation in the brain. The cerebellum as a predictive space-time metric tensor. Neuroscience 7:2949–2970

Pellionisz A, Llinàs R (1985) Tensor network theory of the Metaorganization of functional geometries in the CNS. Neuroscience 16:245–273

Pellionisz A, Peterson BW (1985) Tensor models of primary sensorimotor systems, such as the vestibulo-collic reflex (VCR) and of the metaorganization of hierarchically connected networks. Soc Neurosci Abstr 11:83

Pellionisz A, Baker J, Wickland C and Peterson BW (1986) Tensor network model of the vestibulo-collic reflex in the cat. Exp Brain Res (in preparation)

Peterson BW, Baker J, Wickland C and Pellionisz A (1985) Relation between pulling directions of neck muscles and their activation by the vestibulocollic reflex: tests of a tensorial model. Soc Neurosci Abstr 11:83

Pitts WH, McCulloch WS (1947) How we know universals: The perception of auditory and visual forms. Bull Math Biophys 9:127–147

Reichardt WE, Poggio T (1981) Theoretical approaches in neurobiology. MIT Press, Cambridge

Robinson DA (1982) The use of matrices in analyzing the three-dimensional behavior of the vestibulo-ocular reflex. Biol Cybern 46:53–66

Santini M (ed) (1975) Gogi centennial symposium: Perspectives in neurobiology. Raven Press, New York

Sherrington C (1906) The integrative action of the nervous system. Scribner, New York

Simpson JI (1983) Transformation of coordinates intrinsic to the vestibulo-ocular reflex. Soc Neurosci Abstr 9:315

Simpson JI, Graf W (1985) The selection of reference frames by nature and its investigators. In: Berthoz A, Melvill-Jones J (eds) Adaptive mechanisms in gaze control. Facts and theories. Elsevier, Amsterdam, pp 3–16

Simpson JI, Pellionisz A (1984) The vestibulo-ocular reflex in rabbit, as interpreted using the Moore-Penrose generalized inverse transformation of intrinsic coordinates. Soc Neurosci Abstr 10:909

Simpson JI, Graf W, Leonard C (1981) The coordinate system of visual climbing fibers to the flocculus. In: Fuchs A, Becker W (eds) Progress in oculomotor research. Elsevier, Amsterdam New York, pp 475–484

Soechting JF, Ross B (1984) Psychophysical determination of coordinate representation of human arm orientation. Neuroscience 13(2):595–604

Szentàgothai J (1950) The elementary vestibulo-ocular reflex arc. J Neurophysiol 13:395–407

Wiener N (1948) Cybernetics: or control and communication in the animal and the machnine. MIT Press, Cambridge

Wrede RC (1972) Introduction to vector and tensor analysis. Dover, New York

Neuronic Equations Revisited and Completely Solved

E.R. CAIANIELLO[1]

1 Introduction

Since my first meeting with Cybernetics, it has been crystal clear to me that its true object is the study of "intelligence" (calling it "natural" or "artificial" is misleading and causes only confusion). Can a physicist, with his mind and tools, have a say in such matter? Although today the question easily receives brazenly affirmative answers (I fear that too many problems lay ahead still totally ignored), at the time of my first endeavours the situation stood quite differently. This was lucky, in that little bias could then poison a young, enquiring mind; luckier perhaps than today, when so many claim problems to be "almost solved" and pour down answers, when even questions cannot yet be soundly formulated. For this reason I shall refrain from sweeping statements and limit my discussion to only one of the three basic elements of my model (see below): a set of nonlinear equations that describe the behaviour of a system of coupled binary decision elements in discrete time ("neuronic equations": NE). Their solution is in any case an essential preliminary to that of the remaining parts, "mnemonic equations" and "adiabatic learning hypothesis", which are strongly connected to anatomical, or technological, structural information.

I shall only mention some points which appear to me now, after a quarter century of meditation, of special relevance, summing them up in few sentences:

1. Strategy: the study of neural models, or organization, or complex structured systems (natural languages, social structures ...) needs a common approach as regards functional and relational aspects.
2. Levels of a hierarchical nature are essential in this study; each level requires its specific logic, mathematics, relation to other levels; their discrete quantization is basic for the stability of the system.

[1] Dipartimento di Fisica Teorica e sue metodologie per le Scienze Applicate, University of Naples, Naples, Italy

Brain Theory
Edited by G. Palm and A. Aertsen

3. Quantitative information about populations of levels, also clearly basic, cannot come at the present time from brain research alone (I have turned to natural languages and all sorts of other systems for it, but this is another story).

Many situations are met which are familiar to a physicist: coarse graining, Wilson's renormalization group, spin-like systems are "natural" to both fields. There is however a crucial difference: a system structured into discrete levels needs additional care, as limiting processes may destroy the very structure one tries to understand. Neuronic equations describe a level of activity; their language, with respect to higher levels, is like that of micro- to various scales of "macro"-physical objects. A "neuron" in them is just a decision element; no one-to-one correspondence is assumed with biological neurons. Nevertheless, there is an interesting connection: by enlarging (if and as needed) the number of "neurons" in our equations, their solutions, which I shall exhibit in (formally!) compact and explicit form, exhaust all behaviours one can possibly expect from any nonlinear system presented in discretized form; this includes dynamics, cycles, chaos, transitions, etc.

Most of present day mathematics, as applied to the physical or other sciences, originates historically from the study of the "continuum": gravitation, fields, etc.; it carries with it an underlying notion of "space", be it Descartes's emptiness or the physicist's "vacuum", into which things happen whose description is to be sought by exploring "neighborhoods", through differential, i.e., linear approximations. Higher order, nonlinear terms are added as corrections when the linear description of reality becomes poor.

Everybody knows, of course, that this is done for lack of techniques apt to solve in a general and exact way even trivial nonlinear equations; the cases in which this is possible are known by name in the literature, and mostly belie the guesses that stem from such "linearizing" Weltanschauung. We find thus two concomitant elements: an historical propensity to "think linear"; the hard fact that the only general method available for computation is to fragment a problem into linear pieces.

There is also, however, an ever increasing amount of instances in which nonlinearity is basic: all that goes with boolean algebra, computer sciences, decision-making, models of neural activity, etc.. This is an entirely different sort of universe for the student of Nature, with a correspondingly different *Weltanschauung.* Also here, things become as clumsy (though elegant verbiage may act as a cover) as they frankly appear to a physicist whose expansions refuse to converge. The long-term behaviour, the collective actions of aggregates of interconnected yes-or-no decision elements, and many other such questions, appear unanswerable; or at least, the effort is not made.

Both these aspects are present in the model of neural activity which we proposed (Caianiello 1961) in order to extend and algebrize the pioneering work of McCulloch and Pitts (1943). In it, the behaviour of a net of interconnected yes-or-no elements – the "neurons" – is described by means of three distinct types of laws:

1. those which describe the behavior of the net with constant, or frozen connections: neuronic (or decision) equations (NE);
2. those which accont for the change of the couplings among neurons, i.e. the structure of the net, as a consequence of the activity described by I: mnemonic (or evolution) equations (ME);
3. those that fix the respective time scales appropriate for I and II; ME may require a sequence of them, the smallest of which was estimated for biological neurons, from retrograde amnesia, to be $\sim 10^6$ larger than that of NE. We called this the adiabatic learning hypothesis (ALH); it is clearly necessary to decouple I from II (we recall N. Wiener's comment to us: "without it, we would be playing Alice's game of croquet, changing rules during the game").

The main challenge was presented by part I of the program, NE. The interested reader may find in Ref. [1] a description of the whole model, which we still consider as a sound basis for further study. We restrict our discussion here to a synopsis of results obtained since then on NE, several of which unpublished. We shall comment first on some features of them which appear of special relevance.

The NE express, at time $t + \tau$ (τ a constant delay) the state of the net as determined by its situation at time t. According to the problem at hand, it may be convenient to use values (0.1) or (-1.1) to denote the two allowed states of a "neuron". We need thus to take Heaviside or signum functions of real functions of binary variables: the discontinuous and the continuous aspect appear tied together in an essential way; their interplay is fundamental to NE, whatever the system, neural or not, described by them.

Its full understanding was possible because of a lucky circumstance. Many authors have preferred nonlinearities of smoother types, e.g. sigmoid, or quadratic, etc., so as to use the mathematics of continuum, i.e. standard calculus. It was just our taking the opposite approach, i.e. using consistently totally discontinuous functions, that permitted the exact solution of our NE and all problems connected with them, as will be outlined in the sequel. The breakthrough came about in a most elementary manner, from the obvious property

$$\operatorname{sgn} xy = \operatorname{sgn} x \cdot \operatorname{sgn} y,$$

which may be said to express a full half of the properties of linear functions.

As a conclusion to this introduction, we wish to emphasize that any general model of neural activity, including ours, can only express the laws of a neural medium, not certainly the behaviour of a brain or a robot. Likewise, physics gives laws, not automobiles or TV's: for these, rules have to be found or invented. Our search for appropriate rules (which, at another level, become of course again laws) has led us to different paths: natural languages (Caianiello 1974) and then models of hierarchically structured system (Caianiello 1977); these will not concern us here. The exact solvability of NE puts them on the same footing as linear equations for the study of exact or approximate models of general systems: we shall focus in this report our attention exclusively on this point.

I wish also to note that the application of so crude a model to biological situations (where a real neuron may be conceived as a VLSI made of our "mathematical" neurons) gave results for exceeding our expectations (Caianiello and De Luca 1966, Caianiello and Lauria, 1970).

2 Notation

The "neuron" is a binary decision element, whose states can be better described as $x = (0,1)$, or $\xi = (-1,1)$, according to the specific purpose; of course

$$x = \frac{1+\xi}{2} .$$

The net has N neurons, whose interconnections determine its structure. We are not concerned here with specific structures; the NE describe thus a general net as if it were a physical medium of which the NE describe the laws.

Denote with

$$x \equiv \boldsymbol{x} \equiv \{x^1, x^2, \ldots, x^N\}; \quad x^h = 0,1$$

$$\xi \equiv \boldsymbol{\xi} \equiv \{\xi^1, \xi^2, \ldots, \xi^N\}; \quad \xi^h = \pm 1$$

variables, vectors, or one-col. matrices, whose components have values as specified. Let $F(\xi)$; $\Phi(x)$ be any real functions subject only to the condition

$$F(\xi) \neq 0; \quad \Phi(x) \neq 0$$

for any choice of variables ξ^h, x^h.

This requirement (which is not in fact a restriction) will simplify remarkably our discussion. Call

$$1[\Phi] = \begin{cases} 1 \text{ for } \Phi > 0 \\ 0 \text{ for } \Phi < 0 \end{cases} \quad \text{(Heaviside step function)}$$

$$\sigma[F] \equiv \mathrm{sgn}[F] = \begin{cases} 1 \text{ for } F > 0 \\ -1 \text{ for } F < 0 \end{cases} \quad \text{(signum).}$$

Define
$$\langle F(\xi)\rangle = \frac{1}{2^N} \sum_{(\xi'=\pm 1,\dots \xi^N=\pm 1)} F(\xi^1,\xi^2,\dots,\xi^N) \text{ (trace)}.$$

The tensor powers of ξ have 2^N components:

$$\eta^\alpha = \begin{cases} 1 \equiv \xi^0 \\ \vdots \\ \xi^h & h = 1,2,\dots,N \text{ (linear terms)} \\ \vdots \\ \xi^{h_1}\xi^{h_2}\dots\xi^{h_r} \\ \vdots & \alpha \equiv h_1 h_2,\dots,h_r \text{ (nonlinear terms)} \\ \xi^1\xi^2\dots\xi^N \end{cases}$$

$\eta \equiv \boldsymbol{\eta} \equiv \{\eta^\alpha\}$ is thus a vector in 2^N dimensions, $\eta^\alpha = \pm 1$; the α-ordering of the indices $0,1,\dots,2^N-1$ may be arranged to suit particular needs; we choose here

$$\boldsymbol{\eta} = \begin{pmatrix} 1 \\ \xi^N \end{pmatrix} \times \begin{pmatrix} 1 \\ \xi^{N-1} \end{pmatrix} \times \dots \times \begin{pmatrix} 1 \\ \xi^1 \end{pmatrix} = \begin{pmatrix} 1 \\ \xi^1 \\ \xi^2 \\ \xi^1\xi^2 \\ \xi^3 \\ \vdots \\ \xi^1\dots\xi^N \end{pmatrix}$$

Then, all the properties of boolean functions needed here can be readily derived from the evident ones

$$(\sigma[F])^2 = +1; \quad \sigma(\sigma[F]) = \sigma[F]; \quad \sigma[FG] = \sigma[F]\sigma[G] \ .$$

In particular, one has the η-expansion

$$\sigma[F(\xi)] = \sum_{\alpha=0}^{2^N-1} f_\alpha \eta^\alpha \equiv f^T \eta \ ,$$

where

$$f_\alpha = \langle \eta^\alpha \sigma[F(\xi)]\rangle = \langle \sigma[\eta^\alpha F(\xi)]\rangle \ .$$

If one wishes to use the (0,1) Heaviside functions, posing

$$\chi \equiv \boldsymbol{\chi} \equiv \begin{pmatrix} 1 \\ x^1 \\ x^2 \\ x^1x^2 \\ \vdots \\ x^1x^2\dots x^N \end{pmatrix} ,$$

one finds easily,

$$\chi = C\eta \ ,$$

with

$$C = \begin{pmatrix} 1 & 0 \\ \frac{1}{2} & \frac{1}{2} \end{pmatrix} \times \begin{pmatrix} 1 & 0 \\ \frac{1}{2} & \frac{1}{2} \end{pmatrix} \times \ldots \times \begin{pmatrix} 1 & 0 \\ \frac{1}{2} & \frac{1}{2} \end{pmatrix} ,$$
$$(N \text{ times})$$

so that the connection between the η- and the χ-expansion is given by

$$1(\Phi[\chi]) = g^T\chi = g^TC\eta = \tfrac{1}{2}(1 + \sigma[\Phi]) + \tfrac{1}{2}f^T\eta$$

or, with

$$\tilde{f}_0 = 1 + f_0, \ \tilde{f}_\alpha = f_\alpha \quad (\alpha > 0) \ ,$$

by

$$g = \frac{1}{2}(C^T)^{-1}\tilde{f}; \quad \tilde{f} = 2C^Tg \ .$$

The χ-expansion is less suited than the η-expansion for algebraic manipulations, though more directly related to logic and probabilistic considerations.

3 Linear Separable NE

3.1 The NE

Our first work considered only linear functions $\Phi(\chi)$ or $F(\xi)$, hence l.s. boolean functions $1[\Phi(x)]$ and $\sigma[F(\xi)]$. It is instructive to consider this case first. The NE of (Caianiello 1977) are (we write $u_h \equiv x_h = (0,1)$ for consistency with the notation used there)

$$\text{(I Form)} \qquad u_h(t+\tau) = 1\left[\sum_{\substack{h=1\ldots N \\ r=1\ldots L}} a_{hk}^{(r)} u_k(t - r\tau) - s_h\right] \tag{1}$$

Equation (1) keeps thus into account delayed actions of neurons of the net [$a_{hk}^{(r)} > 0$ excitation, $a_{hk}^{(r)} < 0$ inhibition, $a_{hk}^{(r)}$ loop of self-excitation or inhibition, s_n threshold]. NE written as in Eq. (1), which we may call *First Form*, describe the state of the net at time $t + \tau$: they are *state equations.* They can equivalently be written as *excitation equations,* in the *Second Form*

$$\text{(II Form)} \qquad w_h(t+\tau) = \sum a_{hk}^{(r)} 1[w_k(t - r\tau)] - s_h \ . \tag{2}$$

It can be shown (Caianiello and Grimson 1975) that Eqs. (1 and 2) can be written as well, by enlarging the number of neurons from N to NL, as if without delays [$a_{kh}^{(r)} \to a_{hk}^{(0)} \equiv a_{hk}$]. In matricial notation, setting ($NL \to N$)

$$u_h(m\tau) = u_{h,m}; \quad \boldsymbol{u}_m \equiv \begin{Bmatrix} u_{1,m} \\ u_{2,m} \\ \vdots \\ u_{N,m} \end{Bmatrix}; \quad A \equiv \{a_{hk}\} ,$$

we find

(I Form) $$\boldsymbol{u}_{m+1} = 1[A\boldsymbol{u}_m - \boldsymbol{s}] \tag{3}$$

(II Form) $$\boldsymbol{w}_{m+1} = A1[\boldsymbol{w}_m] - \boldsymbol{s} . \tag{4}$$

3.2 Multivalued Logic

Equations (1) or (3) describe a net with binary decision elements, i.e. 2-valued logic. Equations (2) or (4) can be constructed with suitable A and $\boldsymbol{s}$ so as to give any wanted number k of values at each element (k need not be the same for all elements); *they describe therefore as well a net working with some k-valued logic*; Eqs. (1) or (3) show the connection with binary nets.

3.3 Constants of Motion

Let $\boldsymbol{\gamma}_r$ be vectors, and form from Eq. (4) the scalar products

$$\boldsymbol{\gamma}_r \cdot \boldsymbol{w}_{m+1} = \boldsymbol{\gamma}_r \cdot A1[\boldsymbol{w}_m] - \boldsymbol{\gamma}_r \cdot \boldsymbol{s} .$$

If

$$A^T\gamma = 0$$

we find

$$\boldsymbol{\gamma}_r \cdot \boldsymbol{w}(\tau) = -\boldsymbol{\gamma}_r \cdot \boldsymbol{s} = \text{constant} . \tag{5}$$

If A is of order N and rank R, there are $N - R$ vectors satisfying Eq. (5) and as many linear constants of motion in the net. They can be utilized e.g. as failure detectors; nets may be computed so as to have prescribed constants of motion and no limitation on couplings within rank R. It is also possible, of course, to obtain quadratic constants, etc.

3.4 Self-Dual Nets

We pass now to the signum representation. A particular condition appears then to simplify remarkably the form of Eqs. (3 and 4):

$$AI = 2\boldsymbol{s};$$

this means *self-duality*, and if it holds Eqs. (3 and 4) become $(A \rightarrow \frac{1}{2}A)$:

$$\boldsymbol{u}_{m+1} = \sigma[A\boldsymbol{u}_m] \tag{6}$$

$$w_{m+1} = A\sigma[w_m] \ . \tag{7}$$

Equations (6,7) reduce immediately in turn to the form Eqs. (3,4) by keeping fixed the state of some given neuron ($N-1$ neurons are then free). The self-dual form of N.E. simplifies many computations.

3.5 Polygonal Inequalities

Take $F(\xi)$ linear

$$F(\xi) = \sum_{h=1}^{N} a_h \xi^h, \quad a_n \text{ real numbers}$$

where we assume, to begin with, that

$$a_1 \geqq a_n \geqq \ldots \geqq a_N > 0$$

so that one readily has

$$f_1 \geqq f_2 \geqq \ldots \geqq f_N \geqq 0 \ .$$

These restrictions are easily removed (see later); they serve only to simplify our discussion. We shall be concerned in particular with the relevant case

$$F(\xi) = \sum_{h=1}^{2n+1} \xi^h \tag{8}$$

where of course, $f_1 = f_2 = \ldots = f_N$. Clearly, in Eq. (8) $F(\xi) \neq 0$ always. We also utilize, with $h \equiv (i_1, i_2, \ldots, i_h)$, the notation

$$F^h(\xi) = \sum_{\substack{i \neq i_1, \\ i_2, \ldots, i_h}} a_i \xi^i \ .$$

For self-dual functions the expansion $\sigma[F(\xi)]$ has only odd terms:

$$\sigma\left[\sum_{h=1}^{N} a_h \xi^h\right] = \sum_{h=1}^{N} f_h \xi^h + \sum_{h_1<h_2<h_3}^{1,\ldots,N} f_{h_1 h_2 h_3} \xi^{h_1} \xi^{h_2} \xi^{h_3} + \ldots + f_{1,2,\ldots,N} \xi^1 \xi^2 \ldots \xi^N \ . \tag{9}$$

Consider first

$$\begin{aligned} f_h =& \langle \xi^h \sigma[F(\xi)] \rangle \\ =& \frac{1}{2}(\langle \sigma[a_h + F^{(h)}(\xi)] \rangle + \langle \sigma[a_h - F^{(h)}(\xi)] \rangle) \\ =& \langle \sigma[a_h + F^{(h)}(\xi)] \rangle \end{aligned}$$

hence, the non-zero contributions to f_h come only from configurations $\xi^1, \xi^2, \ldots, \xi^{h-1}, \xi^{h+1}, \ldots, \xi^N$ such that

$$|F^{(h)}(\xi)| < a_h \tag{10}$$

so that

$$f_h = \frac{1}{2^{N-1}} \mathcal{N}_h$$

where $\mathcal{N}_h$ denotes the number of configurations for which Eq. (10) holds. Thus, if

$$a_1 > a_2 + a_3 + \ldots + a_N \tag{11}$$

$\mathcal{N}_1 = 2^{N-1}$, $f_1 = 1$, all other $f_h = f_a = 0$ terms other than the first are irrelevant in Eq. (13). Equation (15) tells that it is impossible to construct a polygon with sides $a_1, a_2, \ldots, a_N$. Otherwise, in general, $\mathcal{N}_h$ is twice the number of distinct triangles which one can construct by taking a_h as base side, and aligning along the two other sides the remaining $a_i \neq a_h$ (their ordering on each of these sides is immaterial) (Caianiello and Simoncelli 1981).

For linearly separable functions the f_h or the numbers $\mathcal{N}_h$ are not independent from one another.

As an illustration (we remind that $a_1 \geq a_2 \geq a_3 > 0$) :

$$\sigma[a_1\xi^1 + a_2\xi^2 + a_3\xi^3] = \frac{1}{2}\xi^1 + \frac{1}{2}\xi^3 - \frac{1}{2}\xi^1\xi^2\xi^3$$

if a_1, a_2, a_3 can be the sides of a triangle,

$$\sigma[a_1\xi^1 + a_2\xi^2 - a_3\xi^3] = \xi^1$$

if $a_1 > a_2 + a_3$. For the extension of this "triangular inequality" to higher N see Caianiello and Simoncelli (1981)

We find that the problem of studying l.s. boolean functions (in particular of determining the coefficients of their η-expansion) proves entirely equivalent to that of computing the numbers $\mathcal{N}_{h_1 \ldots h_{2p+s}}$; the classic triangular inequality is generalized into corresponding appropriate "polygonal inequalities" that pertain to the polygons that can be constructed by assigning at will segments $a_0, a_1, \ldots, a_N$ as sides. This is a fascinating problem of finite mathematics, group theory and topology, which should be investigated on its own merits. Nothing much can be gathered regarding the values $\mathcal{N}_{h_1 \ldots h_x}$ with the customary use of hyperplanes in the study of l.s. functions; it is conceivable that this knowledge may lead to a deeper geometrical understanding of the properties of the $\mathcal{N}_{h_1 \ldots h_k}$ and the yet unknown ways in which these numbers are related to one another.

3.6 Continuous vs. Discontinuous Behaviour: Transitions

We have considered in the text linear arguments $\sum_{h=1}^{N} a_h \xi^h$ with the condition

$$a_1 \geqq a_2 \geqq \dots \geqq a_N > 0$$

that is, the canonical form of a self-dual l.s. function (threshold = 0). All these conditions are most easily removed, as is well known:

1. Putting in the function and in its η-expansion $\xi^h = +1$ or -1 changes a_h into a threshold.
2. A permutation of the variables $\xi^1, \dots, \xi^N$ changes the original η-expansion into that of the permuted ones.
3. Only positivity remains: this is removed by a change of sign of any wanted set of variables, accompanied by a change of $f_{j_1, j_2 \dots j_h}$ into $(-1)^{r_h} f_{j_1 \dots j_h}$, where r_h is the number of indices among $j_1, \dots, j_h$ corresponding to variables ξ^l that have changed sign. To every canonical l.s. function correspond therefore

$$\Omega_N = \frac{2^N N!}{2^Z \Pi s_i!} \begin{cases} z \text{ number of } f_i = 0 \\ s_i \text{ number of } f_j = f_k = \dots = f_i \end{cases}$$

different l.s. functions, whose η-expansions are immediately deducible from the canonical η-expansion by the operations β) and γ) above; thresholds, and even terms, by α).

We find here the most remarkable property of such functions; we can indeed regard the operations β) and γ) as defining different sectors in N space of a same function, α) as a restriction to a semi-space in a given sector. Hence, dynamical behaviour (change of a_h) will consist in jumps, or transitions, across sectors; as long as one stays within a sector, changes are irrelevant. The continuum of real numbers reduces thus, in a perfectly defined way, to the discontinuous behaviour of l.s. functions. Since any boolean function is reducible to a net of l.s. functions, and in the finite and discrete any function is expressible through boolean functions, the claim that already the study of l.s. NE describes the most general nonlinear behaviours is substantiated. How to construct nets to this aim is the subject of Sect. 3.

It is evident to a physicist that the matrix F is the transfer matrix (Q) from a spin state to another (N neurons $\to$ N - spin elements), and that many other such connections can be established. With neural net models we are interested primarily in finding the exact couplings that give wanted behaviours. This is not the type of information one wants when considering spin systems (taken as a paradigm for any discrete, finite, time-quantized model); it is hoped however that when the need arises to characterize – albeit schematically – the subnets of a net as individual elements of a new net, renormalization groups methods may be of use, provided they do not destroy basic structures. This is just an example to say that, once more, the conclusion of our task means only the posing of other ones, in a way that, pleasingly enough, becomes each time more interdisciplinary.

4 The General Case

4.1 η-Expansion

It is convenient here to work directly with the tensorial signum expansions. If each neuron h of the net has as excitation function the real function $f^{(h)} = f^{(h)}(\xi^1, \xi^2, \ldots, \xi^N)$, we can write the NE for a general net as

$$\xi^h_{m+1} = \sigma[f^h(\xi^1_m, \ldots, \xi^N_m)] = \sum_\alpha f^h_\alpha \eta^\alpha_m = f^{hT} \cdot \eta_m \ . \tag{12}$$

We consider now the normalized ξ-state matrix of the net $\phi_{(N)}$; with $N = 3$, e.g., it is

$$\phi_{(3)} = 2^{-3/2} \begin{pmatrix} 1 & -1 & 1 & 1 & -1 & -1 & 1 & -1 \\ 1 & 1 & -1 & 1 & -1 & 1 & -1 & -1 \\ 1 & 1 & 1 & -1 & 1 & -1 & -1 & -1 \end{pmatrix} .$$

We can augment the $N \times 2^N$ ξ-matrix $\phi_{(N)}$ to the $2^N \times 2^N$ η-state matrix, from

$$\eta = \begin{pmatrix} 1 \\ \xi_N \end{pmatrix} \times \ldots \times \begin{pmatrix} 1 \\ \xi_1 \end{pmatrix} ,$$

as follows

$$\Phi_{(N)} = \underbrace{\begin{pmatrix} \frac{1}{2} & \frac{1}{2} \\ \frac{1}{2} & -\frac{1}{2} \end{pmatrix} \times \ldots \times \begin{pmatrix} \frac{1}{2} & \frac{1}{2} \\ \frac{1}{2} & -\frac{1}{2} \end{pmatrix}}_{(N \text{ times})} .$$

$\Phi_{(N)} \equiv \Phi$ is an Hermite matrix such that

$$\Phi = \Phi^T; \quad \Phi^2 = 1; \quad \Phi = \Phi^{-1}$$

$\det(\Phi_N) = (-1)^N$.

We can thus also augment the $N\,\xi - \text{state}\,NE$ to the 2^N η-state form

$$\boldsymbol{\eta}_{m+1} = F \boldsymbol{\eta}_m \ , \tag{13}$$

in which F is a $2^N \times 2^N$ matrix whose first row has all elements $= 1, N$ "linear" rows have the coefficients at r.h.s. of Eq. (9), and the remaining ones are given by tensor multiplication.

We obtain thus the central result that passage from ξ- to η-space linearizes the NE. Thus:

$$\eta_m = F\eta_{m-1} = \ldots = F^m \eta_0 \ .$$

That passage to functional space should linearize the NE is of course not surprising; the relevant feature is that 2^N is (of course) finite, and from now on standard matrix algebra can be used. Clearly, $F\Phi = \Phi P$, P a permutation.

We show next that NE exhibit normal modes, just as linear ones, though more complex than the simple periodic sequences typical of linearity in N-space; they intertwine into "reverberations" (Caianiello 1961), since they stem from linearity in 2^N – space. Their interpretation is in principle the same as that expressed by Eigen and Schuster (1979) for "quasispecies" in their classic discussion of hypercycles (for which NE might be an apt tool).

Let the matrix Δ, $\det(\Delta) \neq 0$, diagonalize F :

$$F\Delta = \Delta\Lambda, \quad \Lambda \text{ diagonal} \ .$$

Then

$$P \cdot \Phi\Delta = \Phi\Delta \cdot \Lambda \ ,$$

i.e. $\Phi\Delta$ diagonalizes P.

Since P is a permutation matrix, its characteristic polynomial (same as of F) is necessarily of type:

$$\lambda^a \prod_b (\lambda^b - 1)^{c_b} = 0 \ , \tag{14}$$

with

$$a + \sum_b bc_b = 2^N, \quad c_b \geq 0 \ .$$

Thus

$b = 1$ implies c_1 *invariant states*

$\lambda = 0$ implies *transients*

$b > 1$ implies c_b *cycles* of period b, corresponding to $\lambda_h = e^{\frac{2\pi hi}{b}}$

(b=1 can of course be regarded as a cycle of period 1).

If we set

$$\eta_m = \Delta\chi_m$$

the NE read

$$\Delta\chi_{m+1} = F\Delta\chi_m = \Delta\Lambda\chi_m$$

so that

$$\chi_{m+1} = \Lambda\chi_m$$

or

$$\chi_{\alpha,m+1} = \lambda_\alpha\chi_{\alpha,m}$$

express the wanted normal modes.

5 Concluding Remarks

5.1 Synthesis

The procedure outlined thus far solves the synthesis problem if what is wanted is a net that leads each state into a prescribed successor: the N "linear" rows of $F = \Phi P \Phi$ determine the coefficients $f_\alpha^{(h)}$ in NE [Eq. (9)]. We are more interested, though, in the synthesis of a net which, starting from a given state, follows in time a prescribed sequence of states; call the wanted sequence

$$\eta_1 = \chi_1;\ \eta_2 = \chi_2;\ \ldots\ ;\eta_{2N} = \chi_{2N}$$

and define F such that

$$\mathcal{F}(\chi_1\chi_2\cdots\chi_{2N}) = (\chi_2\chi_3\cdots\chi_{2N}\chi_1)\ ,$$

that is

$$\mathcal{F}X = X\Theta$$

where the $2^N \times 2^N$ permutation matrix shifts the first column to the last place.

Then

$$\mathcal{F} = X\Theta X_R^{-1} \tag{15}$$

solves the problem. X_R^{-1} means right-inverse, and pseudoinverse matrices will have to be considered whenever necessary (transients); the states χ_α need not be distinct, they will actually reproduce the behavior described by the characteristic Eq. (13) if $\mathcal{F}, X$ are brought into F, Φ by an appropriate permutation. Note that X is *not* normalized (unlike Φ), and that our synthesis problem, through Eq. (15), requires only the knowledge of the "linear" rows of $\mathcal{F}$, which is the product of two known matrices (which, brought into the (0.1)-notation, are sparse).

5.2 Separable and Boolean Nets

Working with permutation matrices P is far simpler than using F. If we request that the net be boolean, it is immediately feasible. If, instead, we restrict the net to be linear separable, or of some other special type, we have to start with F:

$$P = \Phi F \Phi\ ,$$

since the condition for a matrix P to describe a linear separable net is not known (it is actually easy to see that matrices P satisfying a same Eq. (14) may give rise, depending upon the phase relations one chooses for cycles, to separable or to non-separable nets).

Working with this matricial notation, i.e. with η-expansions, does not bring into light a main feature of the NE, the close relation between the continuous and discontinuous aspects of the theory. This point, discussed in (3.6), is essential for any concrete application of the formalism to adaptive or learning devices, or just for reliable design.

We remind, finally, that it is possible in a straightforward manner, by changing number of variables and the value of τ (in a way which we forgo but should be evident in the present formalism) to reduce any boolean function or net to a larger linear net. The consideration of boolean nets cannot yield therefore behaviors that one cannot also obtain from larger linear nets. Use of the one or of other is therefore to be considered as a matter of convenience rather than of principle.

Finally, we note that the analysis of the behavior and synthesis of a net reduces thus to the study of permutation matrices P. The problem is therefore reduced to the study of representations of finite groups generated by the primitive roots of 1, plus degeneracies due to 0's.

References

Caianiello ER (1961) Outline of a theory of thought processes and thinking machines. J Theor Biol 1:209

Caianiello ER (1973) Some remarks on the tensorial linearization of general and l.s. boolean functions. Kybernetik 12:90

Caianiello ER (1974) Programma Procuste per l'analisi di linguaggi naturali. Calcolo 2:83 (1965); Languages, hierarchical structures and logic. Coral Gables Conf "Orbis Scientiae", Jan 1974

Caianiello ER (1977) Some remarks on organization and structure. Biol Cybern 26:151

Caianiello ER, Luca A de (1966) Decision equations for binary systems-application to neuronal behaviour. Kybernistik 3:33

Caianiello ER, Lauria F (1970) Il sistemá nervoso centrale. Atti Conv Med Eur, Ist Angelis

Caianiello ER, Capocelli R (1971) On form and language: The Procrustes algorithm for feature extraction. Kibernetik 8:233; Structural analysis of hierarchical systems. III Int Joint Conf Pattern Recognition, Nov 1976, Coronado

Caianiello ER, Grimson WEL (1975) Synthesis of boolean nets and time behaviour of a general mathematical neuron. Biol Cybern 18:111 (this formalism is connected to the classic results of S.W. Golomb, the so-called Rademaker-Walsh expansion, etc.)

Caianiello ER, Simoncelli G (1981) Polygonal inequalities as a key to neuronic equations. Biol Cybern 41:203

Mcculloch WS, Pitts W (1943) A logical calculus of ideas immanent in nervous activity. Bull Math Biophys 5

Eigen M, Schuster P (1979) The hypercycle – principle of natural self-organization. Springer, Berlin Heidelberg New York

Am I Thinking Assemblies?

C. VON DER MALSBURG[1]

1 What is the Physical Basis of Mind?

The Greeks have reduced the multiplicity of materials to a conceptually simple basis: a small number of atomic types and their chemical combination. Such conceptual unification has yet to be attained for the phenomena of mind. One of the important functions of our mind is the construction of models or "symbols" for external objects and situations. What is to be discussed here is the structure of the symbols of mind.

1.1 Regulative Principles

The discussion may gain in focus by contrast and analogy to the symbols of human communication. The following principles are formulated from this point of view.

1.1.1 Hierarchical Structure

The symbols of communication are hierarchically composed of subsymbols. For instance, a book is composed of chapters, paragraphs, sentences, phrases and words. Such hierarchical structure must also be required of the symbols of mind.

1.1.2 Full Representation

The symbols of communication are mere parsimonious, tokens for the images they are to evoke in the reader's mind. In contrast, the symbols of mind have to fully represent all aspects of our imaginations.

1.1.3 Physical Closure

Each written symbol of communication is represented by a dedicated piece of matter, e.g. a bit of ink on virgin paper. The symbols of mind have to coexist within the same physical system. New symbols should not require new pieces of hardware.

[1] Max-Planck-Institut für Biophysikalische Chemie, 3400 Göttingen, FRG

Brain Theory
Edited by G. Palm and A. Aertsen

1.1.4 A Basis for Organization

The symbols of communication are passive products. In contrast, the symbols of mind have to serve as the basis of active organization.

2 The Classical Framework

Currently, theoretical and experimental work on the function of the brain is dominated by a theory which is so common-place that it is rarely stated as such and that it even lacks a name. Let me refer to it as the "classical framework" or as the "classical theory". Its essential tenets have been stated in the works of Hebb (1949) and Hayek (1952), although most of them are much older.

2.1 Semantics

Physical states of the brain are to be regarded as symbols for objects, situations, etc. What is the structure of these symbols and in what way do they refer to their subject?

2.1.1 Semantic Atoms

The complete symbol characterizing the present state of my mind can be decomposed into smallest units, Semantic Atoms. Each atom can be interpreted as an elementary symbol with its own meaning. Typical subjects for atoms are "a blue line of orientation a and stereo depth r at position x of retina", or "there is a face". Semantic Atoms are represented in the brain by physical units (nerve cells or disjunctive groups of nerve cells). Atoms can be in an active or an inactive state.

2.1.2 Combination by Co-Activation

The complete symbol characterizing the present state of my mind has been composed by the co-activation of a number of semantic atoms. The symbolic meaning of the composite symbol is additively composed of the elementary meanings of the constituent atoms. The sets of co-active units are often called assemblies, a term introduced by Hebb.

2.2 Organization

How are the symbols of mind organized, i.e., how do they come into existence? This organization proceeds in steps which are discussed in reverse historical order.

2.2.1 *Logogenesis*

Units, or atoms, are activated or inactivated by excitatory and inhibitory interactions. These are channeled by slowly changing (see Sect. 2.2.2) physical connections. There are two sources of influences on a unit: external and internal to the brain. The external influences are controlled by stimuli to the sense organs. The internal influences are controlled by the activity of other units. The symbols of mind are quasi-stationary assemblies, stablized by the input and exchange of excitation and inhibition.

2.2.2 *Ontogenesis*

The internal connectivity patterns necessary to stabilize assemblies are created by synaptic plasticity. If an assembly is to be "written" into the system, connections between its active units are strengthened (Hebb plasticity). An assembly thus "stored" can later be recovered from partial input. This function is called associative memory.

2.2.3 *Phylogenesis*

In order for the system to work in the way described it has to be placed in an appropiate initial state, with respect both to the basic machinery and to the initial connectivity. This initial state is realized with the help of information stored in the genes.

2.3 Two Remarks

An accurate and complete statement of the "classical theory" is impossible. As is the case with most fundamental conceptual frameworks, their basic tenets are implicit in all detailed work, yet are subconscious and sometimes inconsistent. The account given here of the "classical theory" therefore is necessarily an idealization.

In the ensuing discussion, the "classical framework" is taken as a comprehensive system for the brain as far as it is concerned with the symbolic representation of the (internal and external) world, although this claim of comprehensiveness is rarely made by anybody.

3 Critique of the Classical Framework

If I ask you to pass me that book over there – and you do so – you demonstrate a number of abilities of your brain: language interpretation, pattern recognition and visual scene analysis, planning of action, visuo-motor coordination and motor pattern generation. It has been tried, over decades, to understand and model those abilities and others within the classical framework, without much success. The problem could be that no one has been able yet to find the right initial connectivity diagrams or the right set of trigger features to make the system work. It is argued here that there is a deeper reason for this failure, that the classical framework itself is to be blamed. Although my dicussion seems to be philosophical in nature, the thrust of it is directed at the solution of technical problems. To put it in computer-language, without appropriate data structures one cannot write appropriate algorithms. This section raises a number of objections to the classical framework. The next chapter proposes solutions to the issues raised.

3.1 Semantics

3.1.1 The Assembly has No Structure

The mental symbol in the classical framework, the assembly, is a set of coactive units. As such it has no internal structure. [I should mention here that the lack of syntactical structure in the assembly has been criticised before by Legendy (1970).] Put the other way around, if two symbols (two sets of units) are made co-active, all information on the partition of units among the original symbols is lost. The classical symbol thus violates principle 1.1.1. It cannot be avoided to co-activate symbols, e.g. those referring to parts of a scene to be described. In such cases the features and modifiers which appertain to the simultaneously represented objects begin to float and form illusory conjunctions, i.e., they trigger consequences which should be reserved for objects combining the features differently. This problem may be called the superposition catastrophe of the classical framework (see Fig. 1).

There are various ways in which the classical framework has tried to avoid this difficulty. The most wide-spread of them consists in a severe restriction of the network to those connections which draw the right consequences. One important scheme is as follows (see Fig. 2). Keep the part-assemblies in phys-

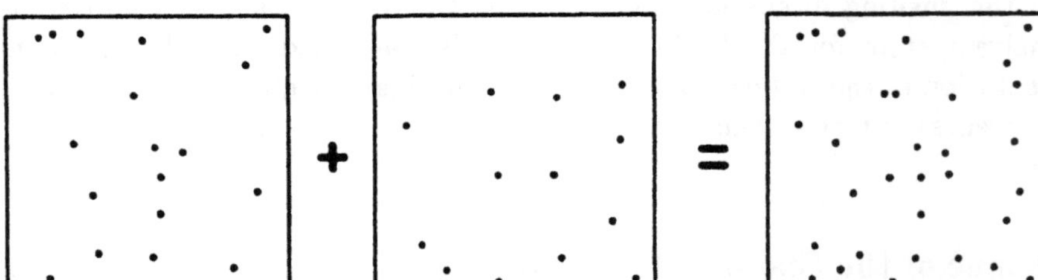

Fig. 1. Superposition catastrophe. The box symbolizes some part of the brain, and is imagined to be filled with neurons. Each *dot* represents an active cell. On the *left,* there are two different sub-assemblies. On the *right,* the two sub-assemblies are superposed by co-activation. Information on the partition of the superposition into sub-assemblies is lost. The rest of the brain can only react to the whole assembly, not to the orginal sub-assemblies. False conjunctions of features will therefore lead to erroneous reactions

ically distinct subregions ("boxes") within the brain (e.g. different parts of a topologically organized visual area); no confusion is possible so far. Have the pattern within a "box" classified by a set of specialized units ("cardinal cells"). Let the "boxes" speak to the rest of the brain exclusively with the help of those "labeled lines". The symbols within the boxes can interact with each other on a higher level without confusion because all detail is invisible on that higher level. Let me illustrate. If the linguistic part of the brain received a full list of the visual features seen at present, it would infer illusory patterns from false conjunctions of features. The presence of patterns has to be evaluated at a place, near to retina, where the relative positions of features are still known, and must be sent in encoded form to the linguistic part. Knowledge of the features is useless if it is not complemented with knowledge about their grouping into patterns. Therefore, visual features are to be hidden from the rest of the brain.

This solution creates more difficulties than it solves. Cutting the network into boxes severely restricts flexibility. It puts a heavy burden on phylogenesis (if it is not possible to derive the box structure from a process of ontogenetic organization). Putting patterns (e.g., visual patterns) into boxes is a difficult problem itself (visual patterns may overlap on the retina!). The scheme presupposes the existence of dedicated units to represent high-level patterns (e.g. "grandmother"). New patterns require new units (thus violating regulative principle

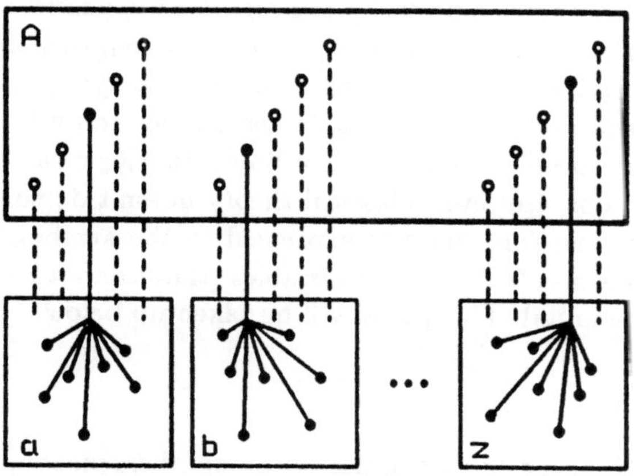

Fig. 2. Avoidance of the superposition catastrophe in the classical framework. Assemblies are enclosed in *boxes, A, a,b,. . .,z.* Active units are represented by *filled circles,* inactive units (in *A*) by *open circles. Patterns* within the lower boxes are represented, within *A,* by cardinal cells (*labeled lines*). Boundaries of boxes are not crossed by connections, except the ones shown. Especially, output from the lower boxes is provided exclusively by cardinal cells. The associative connections necessary to store and stabilize assemblies are restricted to within boxes. The hierarchy can be continued above and below. Patterns in different boxes can be superposed without confusion, since no unit is sensitive to co-activity of units in different boxes. However, the scheme creates great problems. It is difficult to attain one-pattern-per-box. All patterns have to be represented by cardinal cells; information beyond this classification is discarded. The restriction of connectivity as described is difficult to produce (in ontogeny or phylogeny) and is inflexible

1.1.3, physical closure). Awful administrative problems are involved in deciding when to represent a new pattern and in finding virgin units to represent them. All variable detail describing the patterns within the boxes (e.g., "grandmother has a good-humoured grin on her face and wears a blue hat") are stripped off on the way to higher levels, thus violating principle 1.1.2, full representation. In summary, the scheme only works in very restricted environments and with rigid patterns for which the machine has been specifically designed.

Another solution to the superposition catastrophe problem involves selective attention (Treisman and Gelade 1980, Crick 1984). The total symbol which is relevant to the actual situation is not simultaneously active. At a given moment, activity is restricted by a central-command system (directing a "spot-light") to a smaller subsymbol. Conjunctions are only permitted between units which are co-activated in one "fixation" of the spot-light. In this scheme, a history of consecutive fixations can express a symbol which is hierarchically structured into subsymbols, thus solving the problem in principle. Remaining problems with this scheme are with the generation of an appropriate history of activations [the special case of the activation of compact areas in visual space has been worked out in (Crick 1984, Treisman and Gelade 1980)] and with the evaluation of this history. Evaluation necessitates a temporal storage medium which can be sensitive to the whole history, i.e., to a comprehensive symbol.

3.1.2 No Assignment of Meaning to Connections

The classical framework assigns meaning to units. This assignment is possible on the basis of the special contexts in the external and internal environment of the brain in which the units are active. Connections are laid down by synaptic plasticity in response to coincident unit activity. Thus, in comparison to units, connections are tied to much more specific contexts. Accordingly, specific meaning could be assigned to connections. However, classical theory doesn't do so. The reason is that individual connections are not expressed in the symbols (assemblies), and that connections aren't dynamical variables (they cannot be activated and inactivated like the units). This point will be taken up below.

3.2 Logogenesis

Some readers may be inclined to shrug off semantics as an epiphenomenon, insignificant for the dynamics of the brain. However, the points of criticism raised in the last section materialize in terms of dynamics. The important point is that situations which are represented by indistinguishable states cannot be expected to lead to distinguishable consequences in the brain.

3.2.1 Interference by Irrelevant Connections

The point of logogenesis in the classical framework is to co-activate units which are part of the same context. This process is plagued by the presence of connections which have been formed, and make sense exclusively, in presently irrelevant contexts.

In the associative memory scheme of the classical framework, this problem is solved statistically. A unit which should be on in a given assembly receives many excitatory connections from within the assembly, whereas a unit which should be off receives only a few connections, from those other units in the assembly with which it happens to be co-member in another assembly. This system works well as long as the assemblies stored in the memory have little overlap with each other. If, however, assemblies have large subsets in common they start to create strong shadows, partly activating other, overlapping assemblies. Now, overlap between mental symbols is the rule rather than the exception. Traditionally, one avoids the difficulty by introducing more units. If there are several units for each elementary symbol, different copies can be dedicated (together with their connections) to different assemblies. Thus, a given assembly gets rid of irrelevant connections by avoiding the activation of the units that command them.

This "solution" necessarily has an unwanted side-effect. The overlap between mental symbols is important as basis for vital generalizations. When I consider a particular scene, I absorb knowledge about the objects involved, by modifying the interactions within and between the corresponding mental symbols. I want to be able to have this knowledge at my disposal in other situations if they involve partly the same objects or aspects. This, however, is possible only through physical overlap between mental symbols. Avoiding this overlap detroys the basis for generalization.

What is needed is a system in which presently irrelevant connections can be switched off, precisely as the presently irrelevant units can be switched off. This would make connections the subject of "meta-interactions". Such a system is discussed in Sect. 4 below.

3.2.2 Overlay of Functions

This paragraph treats in fact a special aspect of Sect. 3.2.1. Neural modelers usually take the liberty of concentrating at any one time on one function of the nervous system, correspondingly dedicating their hypothetical neural hardware to that function. This is an understandable habit, but it hides the fact that all functions of which our mind is capable have to somehow coexist in the same brain. An obvious solution to the problem is the juxtaposition of dedicated hardware. As far as the functions are already known to phylogeny, this is a viable solution. However, many functions are learnt during ontogeny, and they naturally involve symbols which are partly identical. Functions are defined by appropriate systems of interconnections. While the system is performing one function, connections subserving other functions are highly disturbing. Ontogeny may be able to slowly separate the important functions physically from each other (e.g. by developing connections mediated by specialized units which are gated by excitation and inhibition). However, this will not always be possible (due to anatomical constraints), it will take time (during which an overlay of functions has to be borne), and it may not be desirable (since new functions have to partly use old functions). Again, it would be nice if there was a way to temporarily inactivate all connections subserving functions which presently are not relevant.

3.3 Ontogenesis

The term Ontogenesis is taken here in the special sense of referring to that part of structural genesis of our mind and brain which is shaped by our mental history. Those earlier parts of structural genesis which are under more direct control of the genes, and consequently of evolution, are counted under the heading of phylogeny. Ontogenesis speaks of the formation of memory. For this discussion it is advantageous to distinguish two types of memory, historical and structural memory. A similar distinction is made by many neurologists, although they refer to it by a variety of names. Historical memory allows us to store specific high-level knowledge, such as "Paris is the capital of France", and to recall specific events of our personal biography, complete with detail, circumstances and persons involved, thoughts we had at the time, and so on. Structural memory stores structure and knowledge in a way independent of specific contexts. The two types of memory serve different important purposes. They seem to be implemented by different structures or mechanisms in our brain [it has been shown that patients with amnesia, in whom the ability to lay down historical memory traces is destroyed, may still have structural memory (Cohen and Squire 1980)].

In the classical framework, memory is implemented by Hebb's rule of synaptic plasticity. It connects units which are co-active in a mental state. Let me first discuss structural memory. Certain co-activity relationships in the scene represented by the actual mental symbol are essential, others are accidental. The internal structure of objects is much more stable, and consequently more significant, than, for instance, their spatial relationships, which vary frequently. (This difference in significance of relationships in an observed scene is a direct reflection of the difference in strength of physical interactions within and among the objects making up the scene.) It is usually useless to couple two units which correspond to features which are part of different objects. Such connections, created by the indiscriminate stickiness of Hebb's rule, would soon clutter up the whole brain with unspecific connections.

In classical theory, there are two solutions to this problem. The first relies on the fact that physically non-existent connections cannot be plastically strengthened. It therefore suffices to restrict the physical network to such connections as can be expected to correspond to significant relations, resorting, for instance, to the box structure discussed in Sect. 3.1.1 and in Fig. 2. This solution is to be rejected with the same arguments as used in Sect. 3.1.1. The second solution permits all associations between the units active in a mental symbol, but puts connections under the constraint of competition (e.g. by limiting the total strength of connections going into a unit or coming out of a unit). After sufficient statistics has been gathered, the strengths of connections will reflect the frequency with which they have been strengthened, and thus will reflect the significance of interactions. This solution takes prohibitive amounts of time. In reality we often have to base vital decisions on inspection of a single scene. Somehow, synaptic plasticity must be conditioned by a significance of relations which can be deduced directly from the structure of a scene. Such a system will be discussed below.

Historical memory seems to require the coupling of all units which happen to be united in the mental symbol to be memorized, so that an indiscriminate stickiness is required here. However, historical memory apparently cannot store just any conceivable situation, it seems to be good only for situations which are "legal" according to our structural memory (Bartlett 1932). It may therefore be useful to discriminate a "strong force", which constitutes structural memory, from a "weak force" responsible for historical memory. The weak force becomes perceptible only if the memtal symbol comes sufficiently close to one of the historical memory traces, in which case a large number of individually weak connections coherently add up to perceptible influences. Under ordinary circumstances, the weak force is nothing but an unimportant random perturbation. In any case, the existence of historical memory cannot distract from the conclusion that structural memory cannot consist in an indiscriminate stickiness, as is implied in the classical framework.

3.4 Phylogenesis

Throughout centuries, one of the great themes about the brain has been the balance between nature and nurture, between phylogeny and the more fluid types of organization, ontogeny and logogeny. The genetically determined brain is certainly not a *tabula rasa*. On the other hand, it is to be regarded a weakness of classical theory that for every function it is ready to invoke a specialized connectivity pattern, and that whenever it is difficult to explain it by ontogenetic mechanisms, it invokes genetic determination, i.e. it makes phylogeny responsible for it. This "phylogenetic loophole" is favoured by an important scientific idea, the algorithmic scheme (sometimes referred to by the adjective computational). It proceeds in four sequential steps: (1) identification of a problem; (2) formulation of an algorithm; (3) implementation (in a computer or in the form of a nervous network); (4) execution of the algorithm. If the scheme is applied to specific functions in the brain, it is only too tempting for the modeler to restrict his explicit presentation to steps 3 and 4 and to assign steps 1 to 3 to phylogeny. However, with a problem presenting itself to an individual for the first time in evolution, all four steps have to be performed by ontogeny and logogeny in his brain, and the role of phylogeny must be restricted to providing a "meta-algorithm", the physical framework for the whole scheme, which includes the formulation of the problem and the "invention" of an appropriate "algorithm"! In other terms: we must not only understand how the brain performs certain specific functions, but also how the brain finds them! If it is necessary to invoke specific connectivity structures (cardinal cells, box structures, gated pathways) we also have to specify the process of ontogenetic organization for it. Only structures of a general type can be put off to phylogeny.

3.5 Violation of the Regulative Principles

Assemblies as symbols of the mind violate all of the regulative principles stated in Sect. 1.1.

(1.1.1 Hierarchical Structure) The only subsymbols of an assembly are individual atoms. All information represented by the activity of an assembly can be stated by giving a list of active units (the order in the list having no significance). The absence of any intermediate levels in this "hierarchical structure" is most drastically illustrated by the loss in grouping information when several subassemblies are co-activated (superposition catastrophe, Fig. 1). Several stages of hierarchical structure are only possible if the multi-unit detail of one level is encoded by units (cardinal cells) on the next level.

(1.1.2 Full Representation) To avoid false conjunctions, all specification of intermediate objects in terms of sets of units must be hidden from the view of consecutive stages of processing. Objects are represented by cardinal units, i.e. by mere tokens. In communicating, parsimony forces us to abbreviate complicated structures and convey them by symbolic tokens. In the brain, the full structure of the objects represented should be made available to all subsystems, without being deformed and mutilated by the prejudices of a narrow and rigid coding scheme.

(1.1.3 Physical Closure) One of the striking performances of our brain is its ability to deal with new phenomena and problems. Apparently our mind can build up new symbols and new functional structures. If, as seems to be necessary within the classical framework, patterns and interactions are represented by dedicated units, new units must continuously be actuated. Whereas with communication, where each new letter is a new piece of paper and a new drop of ink, this creates no harm, it cannot be tolerated for the brain. New units cannot simply inherit structure from the sets of units they are to represent and from other patterns with which they overlap. All their connections must be specified from scratch. Moreover, there are terrible administrative problems with the actuation of new units. When is it time to create a new unit: when a new pattern appears for the first time? If not, how do I keep track of multiple occurrences before having a unit deciated? How do I select a candidate unit which happens to have appropriate anatomical connections and which is not yet dedicated? When will I liberate units standing for ephemeral patterns? All of these problems could be avoided if patterns were represented by those units of which they are composed.

(1.1.4 A Basis for Organization!) The classical framework starts with a very simple principle of organization, synaptic plasticity. Certain steps of organization have been successfully described on that basis, among them the formation of feature representing cells and the formation of interconnection patterns to store and stabilize assemblies. However, the assembly as symbol of the mind forces to assume operations and interconnection patterns which are so peculiar that it is difficult to imagine the form of their ontogenetic organization, to say the least.

For all these reasons the classical theory does not deserve the status of a comprehensive framework for a theory of brain function – it cannot be accepted as a description of the physical basis of mind.

4 Natural Representation

My criticism of the classical framework raised above would be incomprehensible if it wasn't complemented by some constructive response. I therefore give here a concise description of a different theoretical framework, which has been described in detail elsehwere (von der Malsburg 1981, Bienenstock 1985, von der Malsburg 1985).

The physical world is hierarchically structured into objects, their arrangements and their parts. There is no need for an object to be "represented" by a new type of "cardinal unit". A coherent object is simply formed by cohesively binding its constituent elements (crystallites, molecules, atoms, elementary particles). The interaction between objects is not mediated by representatives. It takes in fact place as a direct interaction between the constituents. Correspondingly, one could think of a symbol system in which a pattern formed by a set of elementary symbols is represented collectively by – just those elementary symbols! Let me call this a natural representation. High-level symbols in such a system are just large structured masses of elementary symbols, interactions between high-level symbols are the summated effects of the interactions between "atoms". In order for such a system to work one has to introduce degrees of freedom and interactions which allow the atoms to bind to each other in a flexible way. Both the physical world and the systems of visual communication use spatial degrees of freedom to bind elements and form aggregates. If, in the brain, atoms are to be identified with nerve cells, spatial degrees of freedom cannot be used, since nerve cells are immobile.

4.1 Temporal Correlations

Time is divided into two scales, a psychological time-scale (some tenths of a second) which is characteristic of mental processes, and a fast time-scale (some thousandths of a second). Mean unit activity evolves on the psychological time-scale, but the activity fluctuates around this mean on the fast temporal scale. Units bind to each other by correlating their activity fluctuations. A set of units can be bound into a block by synchronizing their fast activity fluctuations. Several such blocks can coexist if their activity is desynchronized relative to each other: this is the solution to the superposition catastrophe. [Legendy has mentioned temporal relations as a solution to the syntax problem already (Legendy 1970).]

Fluctuations arise from an intrinsic instability of units. Correlations arise in sets of units which receive excitation from a common origin or which are synaptically coupled. [That nervous networks can process correlations has been discussed before by Sejnowski (1981).] The useless and trivial state of global correlation is suppressed by an inhibitory system.

As we know, correlations have important consequences for the activation of nerve cells: neurons are coincidence detectors! If two units are desynchronized with each other they cannot cooperate to excite a third unit. If they are synchronized, they can.

4.2 Dynamical Connections

4.2.1 Modulating Connections

Correlations are shaped by connections. If correlations are to represent variable bindings, connection strengths must vary. This function is called synaptic modulation. The excitatory connection between synchronized units is increased in strength, up to a maximum strength which is characteristic of the connection. (The set of maximum strengths for all connections defines the permanent network.) The excitatory connection between two desynchronized units is decreased in strength, down to the value zero. These changes take place on the psychological time-scale. If there are no signals in the two units, the connection slowly sinks back, within times characteristic of short-term memory, to a resting state, in which it conducts with a constant fraction of its maximum strength. [A different system for reducing the physical network to a sparse "skeleton" has been described by Sejnowski (1981).]

4.2.2 Meta-Interactions

Connections interact with themselves and with each other. A connection self-reinforces: the existence of an (excitatory) connection leads to correlation, which in turn strengthens the connection. Connections cooperate: connections between the same source and the same target help each other to synchronize the source with the target, and consequently help each other to grow, if they don't differ too much in length (number of intermediary units). Connections compete with each other as far as they run against the boundary condition of excluded global synchrony. There exists, thus, a system of "meta-interactions", to which was alluded above (Sect. 3.2.1).

4.3 Logogenesis

Classical theory has to work with very restricted permanent connectivity patterns in order to avoid confusion (e.g. patterns are allowed to converse with each other only through their cardinal units). These restricted connectivity patterns have to be formed during ontogenesis or even during phylogenesis. With natural representation, highly specific connection patterns are formed during logogenesis, i.e. on the psychological time-scale. Two constraints contribute to the specificity of those connection patterns: the structure of the permanent network and the rules of the pattern formation process described in Sect. 4.2.2.

4.3.1 Connection Patterns

They are distinguished by sparsity of activated connections (due to the competition between connections) and by optimal cooperation between the surviving connections. Simulation studies (Bienenstock 1985, von der Malsburg 1985) suggest that connection patterns have topological structure, i.e. they can be decomposed into many "neighbourhoods" of directly coupled units which are coupled by sometimes long indirect pathways. (In a random graph any two units have fairly direct connections.)

4.3.2 Projections Between Patterns

Topological connection patterns can be combined, by sparse projections, to form larger topological connection patterns. Thus, rich hierarchies of symbols can be formed. An important special case is constituted by homeomorphic projections, in which two connection patterns of equal inner structure are joined to each other by a one-to-one projection of connections between corresponding units, thus forming a larger connection pattern. Projections between patterns replace the association between the representative units in classical theory.

4.3.3 Restrictions Imposed by Permanent Connectivity

Connection pattern formation is a highly spontaneous process. It has, however, to be influenced by memory. This is possible if the structure of the permanent network systematically favours certain connection patterns. In an extreme case, the permanent network among a given set of units has itself the structure of a connection pattern. In a less radical extreme case, corresponding to the classical associative memory, the permanent network is a superposition of connection patterns. Connectivity dynamics then has to reduce the superposition and recover one of the stored patterns. Unit-wise overlap between the stored patterns does not lead to confusion. In the extreme case, several connection patterns coexist on an identical support of units and can still be selectively activated [this process has been simulated (von der Malsburg (1985)]. In general, permanent connectivity just imposes certain "grammatical" rules on the form of possible connection patterns, leaving great freedom to the patterns formed.

4.3.4 The Evaluation of Correlations

What are the dynamical consequences of the correlation structure of the signals emitted by a given set *a* of units? Suppose the set of units has permanent sparse projections to a number of other sets of units. *a* can excite activity in one of the other sets only if the correlation pattern on *a* fits the structure of connections in that other set, because only then the individual signals sent by *a* can cooperate with the help of inner connections in the other set. In addition, the signals coming from *a* select that connection pattern on the other set of units which fits the connection pattern in *a*. Different correlation structures in *a* thus establish resonance with (and arouse activity in) different sets of units.

4.4 Ontogenesis

Plasticity increases the permanent strength of those connections which are strongly activated. (It is probably necessary to require in addition that new connections be formed between cellular processes over small distances for cells which are strongly correlated.) A compensating process must decrease connections (and finally break them) under appropriate conditions.

This type of plasticity is far from being an indiscriminate stickiness, because two units are permanently connected only after having been bound to each other in a connection pattern. In other words, a new connection is stored only if it is stabilized and validated by existing indirect connections: it is already decided by logogeny whether a particular connection fits the context. In comparison to the classical framework, ontogeny has to carry a much lower organizational burden.

4.5 Semantics

4.5.1 Atoms plus Relations

A sentence cannot be described by an "unstructured sum" of the symbolic meanings of the letters in it (or even of the words in it). The order of the elements establishes relations. These relations are as much carriers of meaning as the elements. (Any message has a standard distribution of letter frequencies if it is long enough, so that information is contained exclusively in the system of neighbourhood relationships!) Natural representation, i.e. the representation of a pattern by its elements, is possible only if relations of the elements within the patterns are represented as well. In the system discussed here for the brain, the general type of relationship can be interpreted as common membership in sub-patterns (where the sub-patterns can overlap). If the whole pattern is the visual description of a scene, units may represent local visual features, and relations bind those features to each other which apply to the same object, to the same part of an object, or to a local neighbourhood. (If the individual feature units are not specific as to retinal position, the patterns of activity and of relations among the units are then position-invariant representatives of visual patterns.) If the whole pattern is a linguistic structure, units may represent phonemes or morphemes and grammatical roles, and relations bind the phonemes or morphemes into higher units, attach grammatical roles to them and assemble those elements into phrases and sentences. Any mental symbol can be further specified by attaching modifying symbols to it. This attachment has to be made precise by specifying to which part of the symbol each modifier applies. The mental symbols which constitute our thoughts are huge systems of cross-referenced active units, all having vague meaning by themselves, creating precision only in their structured ensemble.

4.5.2 Symbols to Whom?

The symbols of communication are sent by one individual and are received by others. Who is the recipient of mental symbols? Let us leave aside the behaviourist answer, that the only recipient is the motor output, i.e. those sub-symbols which form in the motor modality of the brain. We rather have to ask, who is the subject of perception, and, how is the unity of perception established in the brain? It is an ineradicable misconception that the unity of perception has to be established in a separate center, which in addition is often imagined as being of structureless unity itself. This mental archetype leads to infinite regress and to absurdity. Instead, the unity of mind has to be seen as an organic equilibrium among a great multitude of elements. The mental symbols both send and receive at the same time. Signals sent by one sub-symbol are deciphered by other sub-symbols, and the sending symbol can in turn only establish itself, momentarily, if it responds to the messages and questions sent by others. In the state of unity, each subsymbol encodes in its own terms the situation described by the others. This unity is not reached by leaving out detail but by uniting all detail with the help of relations.

References

Bartlett FC (1932) Remembering. Cambridge Univ Press, Cambridge

Bienenstock E (1985) Dynamics of central nervous system. Proc Workshop Dyn Macrosyst, IIASA Laxenburg, Sept 1984, Aubin JP, Sigmund K (eds) Springer, Berlin Heidelberg New York (to be published)

Cohen N, Squire IR (1980) Preserved learning and retention of pattern analyzing skill in amnesia: Dissociation of knowing how and knowing that. Science 210:207–209

Crick F (1984) Function of the thalamic reticular complex: The searchlight hypothesis. Proc Natl Acad Sci USA 81:4568–4590

Hayek FA (1952) The sensory order. An inquiry into the foundations of theoretical psychology. Chicago Univ Press, Chicago

Hebb DO (1949) The organization of behaviour. A neuropsychological theory. Wiley, New York

Legendy CR (1970) The brain and its information trapping device. In: Rose J (ed) Progress in cybernetics, vol I. Gordon and Breach, New York
Malsburg C von der (1981) The correlation theory of brain function. Int Rep 81–2. Dep Neurobiol Max-Planck-Inst Biophys Chem, Göttingen
Malsburg C von der (1985) Nervous structures with dynamical links. Ber Bunsenges Phys Chem 89:703–710
Sejnowski TJ (1981) Skeleton filters in the brain. In: Hinton GE, Anderson JA (eds) Parallel models of associative memory. Lawrence Erlbaum, Hillsdale
Sejnowski TJ, Hinton GE (1985) Separating figure from ground with a Boltzmann machine. In: Arbib MA, Hanson AR (eds) Vision, brain and cooperative computation. MIT Press, Cambridge
Treisman A, Gelade G (1980) A feature integration theory of attention. Cognitive Psychol 12:97–136

Trion Model of Cortical Organization: Toward a Theory of Information Processing and Memory

G.L. SHAW[1], D.J. SILVERMAN[1], and J.C. PEARSON[1,2]

In the spirit of Mountcastle's [1] organizational principle for neocortical function, and strongly motivated by Fisher's [2] model of physical spin systems, we have introduced [3] a new cooperative mathematical model of the cortical column. Our model incorporates an idealized substructure, the trion, which represents a localized group of neurons. The trion model allows for a completely new framework for information processing and associative memory storage and recall: Small networks of trions with highly symmetric interactions are found to yield hundreds to thousands of quasi-stable, periodic firing patterns, MP's, which can evolve from one to another (see Fig. 1). Experience or learning would then modify the interactions (away from the symmetric values) and select out the desired MP's (as in the selection principle of Edelman [4]). Remarkably, we have found that relatively small modifications in trion interaction strengths (away from the symmetric values) via a Hebb-type algorithm [5] will enhance and select out any desired MP. Conceptually this suggests a radically different approach from those information processing models which start at the opposite extreme of a randomly connected neural network with no periodic firing patterns, and then (via Hebb-type modifications [5] in the synaptic interactions) reinforce specific firing patterns. More recently [6], in studying the associative recall properties of the networks we find that, on the average, any of the initial firing configurations rapidly (in 2 to 4 time steps) projects onto an MP. Another exciting feature is that our model includes the known statistical fluctuations in the post-synaptic potentials. These fluctuations are essential for having the huge number of MP's. We believe that these phenomena are of interest to fields of neurophysiology, cellular automata [7] and molecular scale processors [8], as possibly applied to a future generation of computers.

Despite the substantial theoretical efforts and results in modeling neural networks [9], the basis for the tremendous magnitudes of the processing capabilities and the memory storage capacities of mammals remain mysteries. We believe Mountcastle's [1] columnar organizing principle for the functioning of the neocortex will provide a basis for these phenomena and we construct a

[1] Physics Department University of California, Irvine, CA 92717, USA
[2] Present Address: The Rockefeller University, New York, NY 10021, USA

Brain Theory
Edited by G. Palm and A. Aertsen

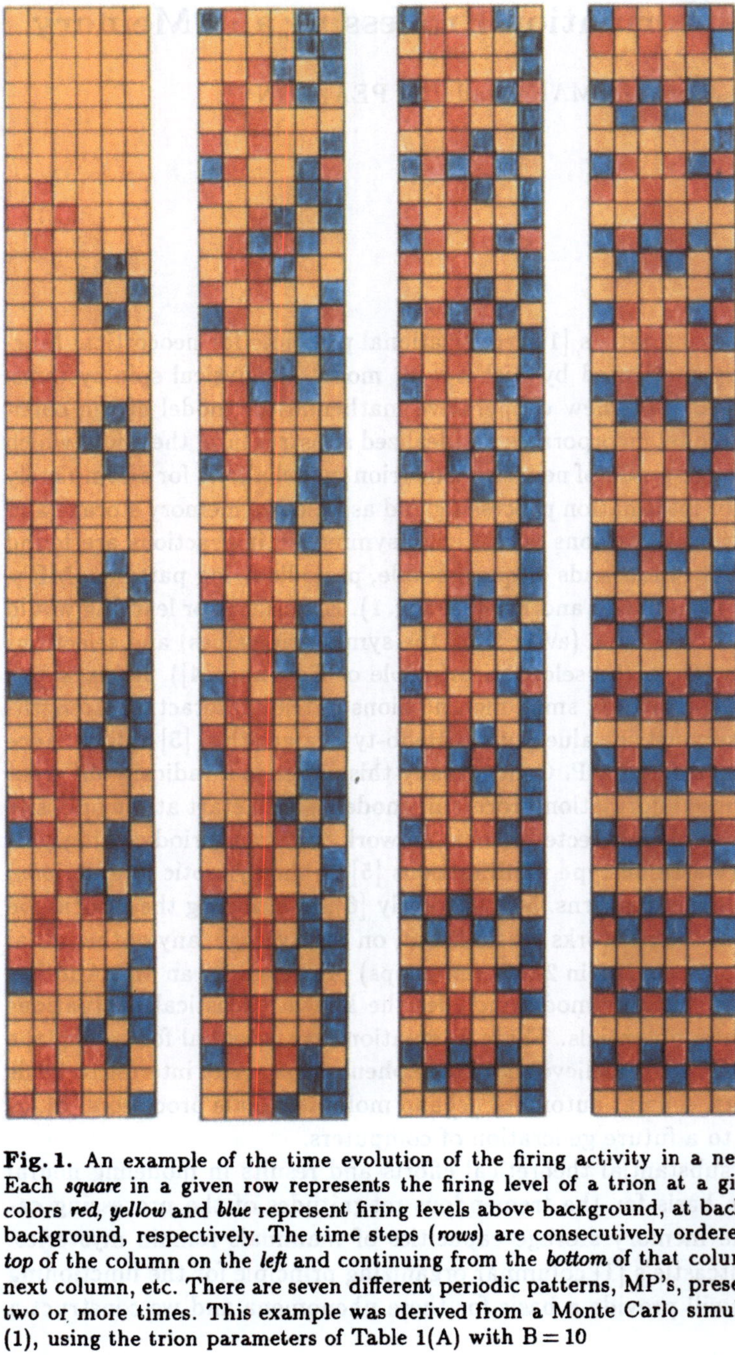

Fig. 1. An example of the time evolution of the firing activity in a network of six trions. Each *square* in a given row represents the firing level of a trion at a given time step. The colors *red, yellow* and *blue* represent firing levels above background, at background, and below background, respectively. The time steps (*rows*) are consecutively ordered, beginning at the *top* of the column on the *left* and continuing from the *bottom* of that column to the *top* of the next column, etc. There are seven different periodic patterns, MP's, present here which cycle two or more times. This example was derived from a Monte Carlo simulation based on Eq. (1), using the trion parameters of Table 1(A) with B = 10

mathematical model based on it. He proposed that the well-established cortical column (roughly 500 microns in diameter) is the basic network in the cortex and comprises small irreducible processing units. The sub-units are connected together into columns or networks having the capability of complex spatial-temporal firing patterns. The creation and transformation of such patterns constitute the basic events of short-term memory and information processing. We strongly emphasize this assumption: that higher, complex mammalian cortical processes involve complex spatial-temporal network neuronal firing patterns; this is in contrast to the usual assumption that the "coding" only involves sets of neurons firing with high frequency. As emphasized by Pearson [10], we are not suggesting that average firing is not important or that it is not a communication code. In fact, we presume that there are several codes in the central nervous system for communication among various regions with the sophistication of the code being related to the sophistication of the information processing involved and to the urgency of the information. For example, the sensing of perilous information must be responded to immediately and presumably would involve a simple alerting code of high neuronal population firing producing a response such as the removal of a hand from a hot stove. In the opposite extreme, the composing (or recall) of a Beethoven symphony must involve incredibly precise, sophisticated spatial-temporal neuronal processes. Most processing of information probably involves several simultaneous types of coding with cortical-cortical coding and processing being more sophisticated than cortical-subcortical. We believe that the key to finding the more complex coding lies in designing multielectrode experiments (in sensory cortex) not only looking at the appropriate spatial and temporal "separations" (we suggest $\sim 50 - 200\,\mu$ and $\sim 30 - 100\,\mathrm{ms}$, respectively) but also presenting the appropriately simple, yet sophisticated stimuli [10–12].

A model presented by Little and developed by Little and Shaw [13] mapped the neural network problem onto a generalization of the (classical) Ising spin model of magnetism. Consider a network of interconnected neurons in which each neuron has two possible states, corresponding to whether or not it has recently fired. These states are updated synchronously in discrete time steps (τ_0), of the order of the refractory period (a few milliseconds, ms.). The state of the system at time $n\tau_0$ is probabilistically related to the state of the system at time $(n-1)\tau_0$ by the interactions between the neurons. The probabilistic feature simulates the known fluctuations in the post-synaptic potentials due to the statistical nature of the release of neurotransmitter. Examination of the solutions of the large fluctuation limit of this model revealed a sub-unit organization [14] in which only a few levels (+, 0, – with respect to background) of the output of each sub-unit of perhaps 30–100 neurons [15] was important[3] The question we addressed [3,6] is, what qualitatively new phenomena can

[3] For example, in a group composed of 60 neurons, firing levels of +, 0, and – could correspond to 60–41, 40–21, and 20–0 neurons firing, respectively. More levels or non-equal and non-symmetric spacing is easily accommodated.

occur as a result of having such spatial substructure and a temporal scale $\sim 25 - 50\,\mathrm{ms}$ as suggested by data from [10–12].

The trion model is an attempt to abstract (in contrast to being derived) from the level of individual neurons to the next level or scale of phenomenological relevance. In making this change of scale we have drawn from the previously discussed work of Mountcastle [1], our theoretical studies [13–15], the ANNNI model work of Fisher [2] and the experiments [10,12]. Fisher showed that a simple extension of the Ising model, the ANNNI (axial next nearest neighbor Ising) model, exhibits a large number of possible stable phases or ordered configurations of spins, whereas the (nearest neighbor) Ising model has just two stable phases. The number of stable phases in these physical models is related to the memory capacity or processing capability in our neural model[4]. There are three essential features [2] responsible for this additional complexity. (1) A strong positive coupling between nearest neighbor spins within each layer which causes them to align, thus providing essential stability against random thermal fluctuations. (2) There are two levels of interaction between spins of different layers: a positive interaction (aligning) between nearest neighbor layers and a negative interaction (reversing) between next nearest layers. (3) Finite fluctuations are essential for the appearance of the large number of phases: in the zero fluctuation limit there are only two stable phases. We have incorporated into our trion model analogs of these three key features and have found that they are all essential for its rich behavior.

In our network composed of interconnected trions, each has three possible states, S, denoted by $+$ $(+1)$, 0, $-$ (-1), which represent a firing output above background, at background, and below background, respectively[3]. Associated with each of the three trion states S is a statistical weighting term $g(S)$ with $g(0) \gg g(-/+)$, which takes into account the number of equivalent firing configurations of the trion's internal neuronal constituents [14,15]. Thus by effectively averaging over the states of its constituents and by using the approximation of only three distinct firing levels with $g(0) \gg g(-/+)$ the output of the trion gains crucial stability, which is analogous to feature 1 above in the ANNNI model. The trion states are updated synchronously in discrete time steps τ $(> \tau_0)$. The state of the system at time $n\tau$ is probabilistically related to the states of the system at times $(n-1)\tau$ and $(n-2)\tau$ by the values of $g(S)$ and the temporal interactions between trions. The dependence on the two previous time steps is analogous to feature 2 above in the ANNNI model, and the probabilistic factor is due to the random nature of synaptic transmission as well as other sources of noise. The probability $P_i(S)$ of the i-th trion attaining state S at time $n\tau$, is given by:

[4] Figure 1 in Reference 2 (Fisher and Selke 1980) illustrates the infinite number of distinct spatially modulated phases in the ANNNI model. None of these phases are co-existent for a given set of parameters (except at a special multiphase point or line), whereas many of the analog quasi-stable periodic firing patterns (MP's) can be excited for a given set of system parameters.

$$P_i(S) = \frac{g(S) \cdot \exp(B \cdot M_i \cdot S)}{\sum_s g(s) \cdot \exp(B \cdot M_i \cdot s)}, \quad M_i = \sum_j (V_{ij} \cdot S_j' + W_{ij} \cdot S'') - V_i^T \tag{1}$$

where S_j' and S_j'' are the states of the j-th trion at times $(n-1)\tau$ and $(n-2)\tau$ respectively. V_{ij} and W_{ij} are the interactions between trions i and j between time $n\tau$ and times $(n-1)\tau$ and $(n-2)\tau$ respectively. V_i^T is an effective firing threshold. B is inversely proportional to the level of noise, "temperature", or random fluctuations in the system. The deterministic limit is taken by letting B approach infinity, (analogous to the noise approaching zero), in which case the $S = 0$ states vanish.

Although our basic formalism, Eq. (1), is not derived from models of single neurons, the simplifying form readily follows from our previous work [14] and the approximation that each trion corresponds to a group of 30 to 100 neurons [15] for which $g(0)/g(-/+) \sim 500$. The key assumption concerning *two discrete* time steps τ clearly warrants further discussion: In analogy with the next nearest neighbor spatial interactions in Fisher's spin model being a crucial feature in obtaining many stable phases, we find that interactions lasting two time steps led to a striking increase in the *number* and complexity of our quasi-stable periodic firing MP's[5]. First we note that the trion time step τ we have in mind is ~ 50 ms, a much larger time than the individual neuronal firing time of a few ms. To establish the *plausibility* of such a τ, we note the observation of periodic bursting in cortex has a long history [16], see, in particular, Morrell [12] who found multipeak responses in cat visual cortex with peak separation of approximately 50 ms. In addition, Morrell [12] observed dramatic changes in these bursting patterns when he paired stimuli in conditioning experiments. Some of these data are shown in Fig. 2. We suggest that these burst pattern data might be consistent with exciting or enhancing different MP's in our trion model. In addition to the simulation studies reported in this paper, we have conducted neurophysiological experiments [10,11] to test certain assumptions of the model. We present in Fig. 3 some of our data from cat primary visual cortex recording from a group of 2–3 neurons which show four equally spaced peaks in the post-stimulus histogram in response to a flashed bar (also see Fig. 1 of [11]). These peaks are separated by approximately 50 ms in close agreement with Morell's data [12]. Also, as a result of showing time sequences of different bar orientations we observed [10,11] burst patterns which might be consistent with exciting MP's. Clearly, it would be very interesting to record simultaneously from two or more closely spaced microelectrodes to test our assumption of a discrete time step τ. Although we have no evidence for interactions lasting two time steps we are able to construct several different mechanisms for it. We believe these are both reasonable and testable. For example, we might imagine

[5] Mathematically, the presence of interactions spanning 2 time steps leads to the description of the system by second order difference equations which are known to yield oscillatory solutions, while 1st order equations only yield exponential solutions.

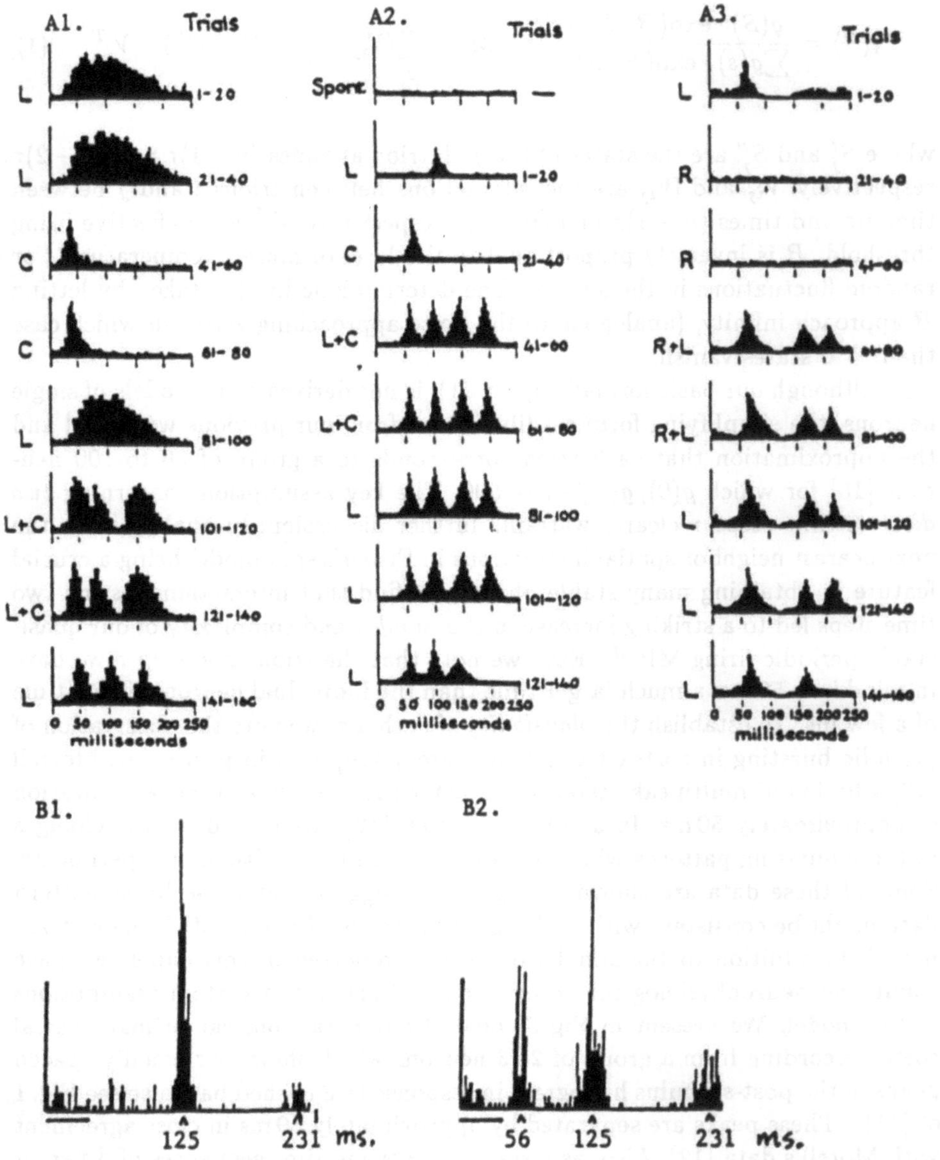

Fig. 2A and B. Single-unit data from visual area III of cat from the published work of Morrell. **A1**, **A2** and **A3** are derived from Figs. 11, 14 and 12 respectively of Ref. [12] (1967). **B1** and **B2** are derived from Fig. 2 of reference [12] (1980). **A** All stimuli were presented during the 0 to 50 ms interval. In **A1** and **A2** *L* denotes a light line, *C* denotes an auditory click, and $L+C$ denotes simultaneous presentation of the light line and the click. In **A3** the same light line was presented to the left eye *L*, the right eye *R*, and to both eyes $R+L$. In **A2** spont. denotes the spontaneous or background level of discharge. The data displayed in each histogram were acquired in consecutive sets of 20 trials, as indicated. The calibration bar at time 0 equals 20 spikes. **B** The visual stimulus was a light line presented during the time marked below the axis. The calibration bar indicates 20 spikes and about 150 trials were given. In **B1** the light line was vertical while in **B2** it was horizontal. These data demonstrate possible complex coding occuring at burst intervals of roughly 50 ms with burst levels of large, small or no peaks

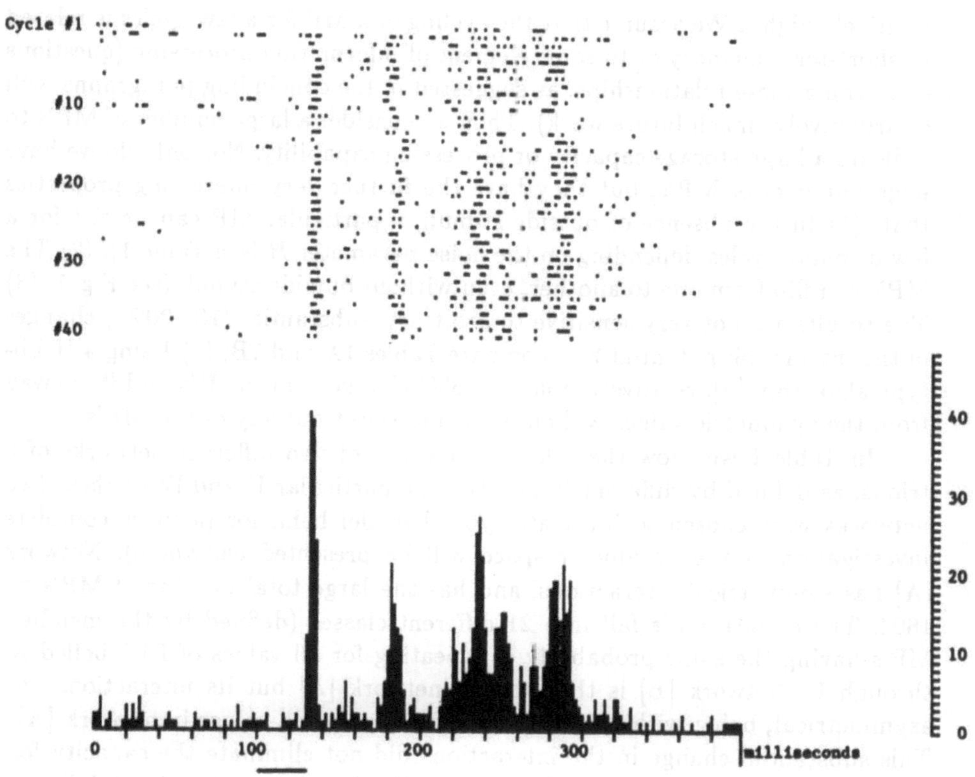

Fig. 3. Spike firing response of cluster of three neurons in area 17 of a cat to a stationary flashed bar (2 1/2° by 1/4°). Forty stimuli were presented with a 400 ms inter-stimulus interval and a 34 ms stimulus duration, and the data is displayed using a 2 ms bin. Note the presence of 4 peaks in the histogram. The interpeak interval is remarkably consistent at 50 ms ± 3 ms, which suggests a trion inter-burst interval of ~ 50 ms. These data are similar to those presented in Ref. [11]

that the upper and lower cortical layers in a column separately are trion networks. If the firing of the upper layers was delayed from the bottom layers by $\tau/2$ (i.e., the peak burst firings were interleaved), then the interaction of the two networks with interactions lasting only one step is readily mapped onto our two time step model. The experimental search for such an interleaving of bursts between upper and lower layers would be of great interest.

We have studied properties of the stochastic time evolution of the states of the system as a function of B, $g(S)$, V and W as given by Eq. (1). Our calculations have considered terms V_{ii}, $V_{ii\pm1}$, W_{ii}, $W_{ii\pm1}$, $W_{ii\pm2}$. The principal finding has been that networks of a small number of trions with highly symmetric V's and W's can exhibit a huge number of quasi-stable, periodic firing patterns. Since any of these patterns can be enhanced and selected out by small changes in the V's and W's (away from the symmetric values) using a standard Hebb-type algorithm [5], we denote these patterns as MP's (magic patterns). Quasi-stable means that the probability of the MP repeating itself or cycling is relatively high[6]. We assume that the cycling of a MP for a few cycles is related to short-term memory or to some element of information processing (questions concerning these relationships, as suggested in the concluding paragraphs, will clearly involve much future work). Thus we consider a large number of MP's to indicate a large storage capacity or processing capability. Not only do we have huge numbers of MP's, but they have the further very interesting properties that: (1) In the absence of outside stimuli, a particular MP can persist for a few or many cycles depending on the noise parameter B (see Table 1); (2) The MP's can flip from one to another, even without outside stimuli (see Fig. 1; (3) The results are not very sensitive to arbitrary, substantial (10–20%) changes in the interaction potential V's, compare Tables 1A and 1B; (4) Using a Hebb-type algorithm [5], relatively small ($\lesssim 5\%$) changes[7] in the V's and W's away from the symmetric values will enhance and select out any of the MP's.

In Table 1 we show the salient properties of two different networks of 6 trions, as defined by different V and W. The particular V and W of these two networks were chosen to illustrate typical model behavior (a more complete investigation of the parameter space will be presented elsewhere). Network [A] has symmetrical interactions, and has the large total number of MP's of 1804. These 1804 MP's fall into 21 different classes (defined by the member MP's having the same probability of repeating for all values of B) labelled A through U. Network [B] is the same as network [A] but its interactions are asymmetrical, being within 20% of their corresponding values in network [A]. This substantial change in the interactions did not eliminate the capacity for large numbers of MP's nor introduce new MP's, thus illustrating the stability of the system. However, breaking the symmetry of the interactions did transform the structure and properties of the MP classes, creating greater diversity of response. Many of the classes in network [A] split to form several classes in network [B], as denoted by subscripts in the class labels in network [B]. The classes also show more interesting variability with the level of noise, especially

[6] The MP's are found by computing the most probable temporal evolution of the trion states from each of the possible initial conditions, and determining if that evolution leads back to the initial conditions with a high probability. Networks of 6 trions with interactions spanning 2 time steps have $3^{(6+6)} = 531,441$ possible initial conditions (the number of configurations of the S's of the first two time steps).

[7] Here we illustrate with a specific example how the Hebbian plasticity algorithm of Eq. (2) enhances the MP's probabilities of cycling even for the "complex" six cycles in a finite range of B. The sample MP is that labelled E_3 in Table 1 [C]. For instance, the changes in the nearestneighbor interactions for the third trion are $\Delta V_{34} = 0.1$, $\Delta V_{32} = -0.02$, derived from Eq. (2) with $\varepsilon = 0.02$. The resulting new probabilities of cycling (using the complete set of changes in the V's and W's including next nearest neighbors for all six trions) are 0%, 24%, 51%, 27%, 0% for B values of 20, 10, 8, 6, 4 respectively, as compared with the corresponding probabilities of Table 1 [A] which are 96%, 80%, 28%, 0%, 0%, demonstrating the significant increase or enhancement of cycling probability due to the Hebbian change for B values 8 and 6. In contrast, these same coupling changes significantly reduced for all values of B the cycling probability of the MP labelled G in Table 1 [C]. Its cycling probability fell from 96%, 67%, 8% to 0%, 27%, 4% for B values of 20, 10, 8, respectively.

the rapid decrease in the probability of repeating at low as well as at high levels of noise. The low noise extinction is because finite levels of noise are needed to sustain the trion $S = 0$ firing levels in asymmetrical networks [as can be seen from Eq. (1), $S = \pm 1$ will dominate as the noise goes to zero]. This suggests that non-specific inputs to the trion network might simulate noise which could tune the network's response properties, i.e., changes in B will change which classes have the highest probability of repeating.

We have found, in general, as illustrated in the example in Table 1 [A], that highly symmetrical trion interactions lead to huge numbers of MP's. We speculate that these symmetrical interactions might be specified genetically giving a "naive" network which could initially respond to many different input signals. Experience or learning (during a critical period [17]) could then modify the connections via a Hebb type mechanism [5] to select out the appropriate responses or MP's. We assume that the cycling of the trion network through a firing pattern will produce small changes in the V's and W's given by the Hebb-type algorithm [5,13]

$$\begin{aligned} \Delta V_{ij} &= \varepsilon \sum_{\text{cycle}} S_i(\tau) S_j(\tau - 1) \\ \Delta W_{ij} &= \varepsilon \sum_{\text{cycle}} S_i(\tau) S_j(\tau - 2), \quad \varepsilon > 0 \ . \end{aligned} \tag{2}$$

By changing the V's and W's away from their symmetric values using Eq. (2), we find that any of the MP's can be enhanced and selected out (in the sense that the particular MP and some others in its class, see Table 1, have their firing probabilities enhanced, whereas most others are suppressed). Although, it is easy to show that Eq. (2) must enhance any one cycle, it seems remarkable to us that this works for every MP in Table 1 (and other examples of MP's we have investigated), even for the very complex six cycles shown in Table 1C (for B values < 10).[7] We believe that our model might provide a specific framework for Edelman's Group-Selective Theory of brain function [4]. The remarkable and rapid reorganization of somatosensory cortex found by Merzenich [18] lends support to Edelman's ideas.

We have carried out the same analysis on networks identical to [A] and [B] except that the statistical weight of the trion zero firing level, $g(0)$, is set equal to zero. This is equivalent to a single neuron model (i.e., 1 neuron per trion), in which there are no $S = 0$ states. These networks are capable of supporting only 12 and 5 MP's respectively. In general, we have found that only networks for which $g(0)/g(-/+) \gg 1$ have large numbers of MP's. In addition to this condition on the statistical weights and the necessity of finite fluctuations, large numbers of MP's require that the interactions span two time steps (i.e., some non-zero W) with both positive (excitatory) and negative (inhibitory) interactions[5].

We have investigated several other aspects of the model (we are exploiting the content-addressable nature of the MP's in our computer studies of these properties). Networks of 7 and 8 trions yield qualitatively similar numbers of

Table 1. Properties of two representative networks with 6 trions. The periodic firing patterns (MP's) are grouped into classes defined by the member MP's having the same probability of cycling for all values of B. In (A) and (B) we list for each network: their defining parameters (the non-zero V, W, and g(+,0,-1); the number of MP's in each class; the percent probability of each MP cycling or repeating for B ranging from B = 4 (high noise level) to B = 40 (low noise level); the total number of MP's represented in the table. By breaking the symmetry of the interactions, 8 of the classes in (A) split into several Classes in (B). This is denoted by subscripting the labels of the split classes in (B). The interactions between the trions in these networks have periodic boundary conditions, which is equivalent to a ring-like spatial connection of the trions. Only those MP's with a cycling probability >10% at B = 10 are included, and probabilities of cycling less than one percent are not listed. In (C) we show a representative MP from each of the classes listed in (A) and (B). We note the interesting diamond like structure in some of these MP's which shows the similarity between M and N and between I_1 and I_2

A

$V_{ii-1} = 1.0$ $V_{ii+1} = 1.0$ $W_{ii-2} = -1.0$ $W_{ii+2} = -1.0$

$g(-) = 1$ $g(0) = 500$ $g(+) = 1$

MP Class	No. of MP's	Fluctuation parameter B: 40	20	15	10	8	7	6	5	4
		% Probability of cycling								
A	17	95	95	95	95	95	94	89	56	2
B	72	94	94	94	94	94	93	88	60	4
C	2	91	91	91	91	91	90	88	69	14
D	2	98	98	98	98	98	97	94	75	15
E	156	96	96	96	80	28	5			
F	144	95	95	95	80	28	5			

B

$V_{ii-1} = 0.8$ $V_{ii+1} = 1.0$ $W_{ii-2} = -1.15$ $W_{ii+2} = -1.1$

$g(-) = 1$ $g(0) = 500$ $g(+) = 1$

MP Class	No. of MP's	Fluctuation parameter B: 40	20	15	10	8	7	6	5	4
		% Probability of cycling								
A_1	5				14	43	57	62	36	1
A_2	12				32	60	70	71	43	1
B_1	36				28	56	66	69	44	2
B_2	36			16	65	78	82	80	53	4
C	2				34	61	70	73	56	9
D	2				37	65	75	79	60	10
E_1	36				18	5	1			
E_2	6			14	28	4	1			
E_3	36		7	44	42	7	1			
E_4	6	54	91	88	19	1				
F_1	36				10	3				
F_2	36			9	24	4				

Table 1 (cont.)

G	432	96	96	96	67	8				
H	216	94	94	94	78	27	5			
I	12	91	91	91	76	26	5			
J	216	95	95	94	79	27	5			
K	51	92	92	92	92	92	92	88	64	8
L	15	98	98	98	98	98	98	96	82	28
M	6	89	89	89	89	89	89	87	75	26
N	6	96	96	96	96	96	96	94	80	28
O	12	98	98	98	82	28	5			
P	12	92	92	91	77	26	5			
Q	210	97	97	97	68	8				
R	6	99	99	99	99	99	99	98	91	53
S	6	97	97	97	97	97	97	96	89	52
T	1	98	98	98	98	98	98	98	98	98
U	210	95	95	95	67	8				

Total number of MP's = 1804

F_3	36		19	54	17	1				
G	108			10	13	4				
H_1	108			4	21	4				
H_2	108		1	23	14	1				
I_1	6			13	26	4				
I_2	6	50	85	82	18	1				
J	108			19	36	7	1			
K	51			7	56	73	78	77	55	6
L	15		35	72	91	94	95	94	80	28
M	6		1	24	70	79	82	82	69	22
N	6		1	26	75	85	88	88	74	24
O	6		35	71	48	8	1			
P	6		33	66	45	7	1			
Q	51		12	51	23	1				
R	6		59	85	95	97	97	98	90	53
S	6		58	83	93	95	95	95	87	52
T	1	98	98	98	98	98	98	98	98	98

Total number of MP's = 883

Table 1 (cont.)

C

```
              Trion
              firing
           ┌-----------
    T s    ¦
    i t    ¦
    m e    ¦
    e p    ¦
```

A_1	A_2	B_1	B_2	C
- - - - - -	- + - - - 0	- 0 - - - 0	0 + 0 - - 0	- 0 - 0 - 0
- - - - - -	+ - 0 - - -	0 - - - - -	+ 0 0 - - -	0 - 0 - 0 -
0 0 0 0 0 0	0 + 0 - 0 0	0 0 0 - 0 0	0 + 0 - - 0	0 0 0 0 0 0
+ + + + + +	+ + 0 + - +	+ + 0 + 0 +	+ + 0 0 - 0	0 + 0 + 0 +
+ + + + + +	+ + + - + 0	+ + + 0 + +	+ + + - 0 0	+ 0 + 0 + 0
0 0 0 0 0 0	+ 0 0 0 - 0	+ 0 0 0 0 0	+ + 0 0 - 0	0 0 0 0 0 0

D	E_1	E_2	E_3	E_4
0 0 0 0 0 0	- 0 - - - 0	+ 0 + - 0 -	- 0 + - 0 0	- + 0 + - 0
+ - + - + -	- - - - - -	0 + + 0 - -	+ 0 - + - -	0 + + 0 - -
- + - + - +	0 + 0 - 0 +	- + 0 + - 0	- + + - 0 +	+ 0 + - 0 -
	+ + 0 + 0 +		+ 0 0 + 0 -	
	+ + + - + +		- + + - 0 +	
	+ 0 - 0 - 0		+ 0 - + - -	

F_1	F_2	F_3	G	H_1
- 0 - - - +	- + 0 - - 0	0 + - - - 0	- 0 0 - - 0	0 0 0 - - +
0 - - - - -	+ 0 0 - - -	+ 0 0 - - -	+ 0 - - - -	0 0 - - - -
0 - 0 - 0 0	0 + + - - 0	+ + 0 - - 0	0 + + - - +	0 - 0 - - 0
+ + - + 0 +	+ + 0 + - 0	+ + 0 0 - +	+ + 0 + 0 0	+ + - 0 0 0
+ + + 0 + +	+ + + - 0 0	+ + + - 0 0	+ + + - 0 +	+ + + 0 0 +
+ 0 0 0 + 0	+ + 0 0 - -	+ + 0 - - 0	+ + - 0 - -	+ + 0 0 + 0

H_2	I_1	I_2	J	K
0 0 - - - 0	- 0 0 0 - 0	0 0 - 0 - 0	0 0 0 - - 0	0 0 0 - - 0
0 0 - - - -	0 0 0 - 0 -	0 0 0 - 0 -	+ 0 - - - -	0 0 - - - -
+ 0 0 - - 0	0 0 + 0 - 0	+ 0 0 0 - 0	0 + 0 - - +	0 0 0 - - 0
+ + 0 0 0 +	0 + 0 + 0 0	0 + 0 0 0 +	+ + 0 0 0 0	+ + 0 0 0 0
+ + + 0 0 +	+ 0 + 0 0 0	+ 0 + 0 0 0	+ + + - 0 +	+ + + 0 0 +
+ + 0 - 0 0	0 + 0 0 0 -	0 + 0 - 0 0	+ + - 0 - 0	+ + 0 0 0 0

Table 1 (cont.)

L	M	N	O	P
0 + + - - 0	0 0 0 0 - 0	0 + 0 0 - 0	+ + + - - -	0 0 0 0 - 0
+ + 0 0 - -	0 0 0 - 0 -	+ 0 + - 0 -	+ + + - - -	0 + 0 - 0 -
	0 0 0 0 - 0	0 + 0 0 - 0	0 + 0 0 - 0	+ 0 + 0 - 0
	0 + 0 0 0 0			0 + 0 0 0 0
	+ 0 + 0 0 0			+ 0 + 0 - 0
	0 + 0 0 0 0			0 + 0 - 0 -

Q	R	S	T	U
0 0 0 - - 0	0 + + 0 - -	0 0 + 0 - 0	0 0 0 0 0 0	0 0 - - - +
+ + - - - -		0 + 0 0 0 -		0 0 - - - -
+ + + - - +				+ - 0 - - 0
+ + 0 0 0 0				+ + - 0 0 +
+ + + - - +				+ + + 0 0 +
+ + - - - -				+ + 0 - + 0

MP's. The particular cyclic boundary conditions[8] used do not appear to be a critical factor. Networks with fixed boundary conditions also support many MP's, as do networks with $g(+)/g(-)$ ratios not equal to one[3]. Monte Carlo simulations of the temporal evolution of trion states, using Eq. (1) and a random number generator, are being employed to study the dynamic properties of transitions between MP's (see Fig. 1) and to study the effect of relaxing our assumption of synchronous firing. Introducing partial asynchrony (by updating at random one trion prior to the others) does not seriously degrade the quasi-stability of the MP's, through some reasonable semblance of a basic time step is necessary to fully exploit the model's potential.

Major fundamental questions remain to be investigated: In working toward a theory of associative memory storage and recall, we need to know why reinforcing any particular MP via the Hebb algorithm, Eq. (2), will enhance it and select it out. This seems truly remarkable for some of the complex, long-cycle MP's. Clearly we need to know how outside stimuli excite the various MP's. Along this line, we have found that in studying the associative recall properties of the networks that, on the average, any initial firing configuration (for example, for a six trion network, we examine all 3^{12} firing configurations of the first two time steps) rapidly (on the average, in 2 to 4 time steps) projects onto an MP. As noted, certain MP's spontaneously flip to another MP's. This

[8] Recent models of the spatial organization of orientation mini-columns in the primary visual cortex of cats and monkeys suggest that they are arranged in circular groupings. Such a spatial organization would naturally lead to ring-like connections and thus to periodic boundary conditions [Braitenberg V and Braitenberg C (1979) Biol Cyb 33:179–187]. Also see Fig. 11 of [Suga N (1978) Fed Proc 37:2342–2354] which shows the circular arrangement of amplitude tuned cell columns in the auditory cortex of the mustache bat.

phenomenon should be relevant in obtaining and understanding hierarchies of associations, as well as the profound problem of obtaining long complex time sequences. Periodic driving inputs and the coupling of two or more trion networks should be considered. The nature of the internal processing within each trion must also be understood before a theory of information processing can be complete.

We conclude that the trion model allows for a completely new framework toward developing theories for information processing and for associative memory storage and recall. Even at these initial stages of the model, we believe that it will be very powerful in simulating multi-electrode recordings in cortex, thus stimulating new ideas in designing and analyzing these important experiments. Verzeano [19] in his pioneering work found evidence for circulation of bursts of spontaneous activity (in thalamus and cortex of awake and sleeping cat) by monitoring spike activity with three or more closely spaced microelectrodes (see, e.g., Fig. 4 of Ref. [19] (1960)). We believe that it will be crucial to both have several closely spaced ($\sim 50 - 200\,\mu$ separation) microelectrodes that can monitor neurons in several nearby trions and have stimuli of the sequential type used in Ref. [10] and Ref. [11] to excite the MP's. If indeed there is an approximately synchronous time step $\tau(\sim 30 - 100\,\mathrm{ms})$ during which groups of cortical neurons tend to burst (see Figs. 2 and 3), this should be exploited in the presentation of stimuli to the animal. Repetitive, dynamic sequences spaced at time intervals τ might resonantly excite the spatial-temporal MP's. It also appears potentially very useful to do conditioning experiments of the type used by Morrell [12] to excite these MP's.

Acknowledgements. We thank M.E. Fisher, M.M. Merzenich, J. Patera, A.B. Scheibel, J.C. Ward, and N.M. Weinberger for many helpful discussions, suggestions and encouragement. We also thank J. Patera and R. Vasudevan for their help and participation in the early stages of the development of this work. We thank the Aspen Center for Physics where part of this research was done. This work was supported in part by the UCI Focused Research Program in Brain Function.

References

1. Mountcastle VB (1978) In: Edelman GM, Mountcastle VB (eds) The mindful brain. MIT Press, Cambridge, pp 1–50
2. Fisher ME, Selke W (1980) Phys Rev Lett 44:1502–1505; (1981) Philos Trans R Soc (London) 302:1–44; (1981) J Appl Phys 52:2014–2018;
   Huse DA, Fisher ME, Yeomans JM (1981) Phys Rev B 23:180–185
3. Shaw GL, Silverman DJ, Pearson JC (1985) Proc Natl Acad Sci USA 82:2364–2368
4. Edelman GM (1978) In: Edelman GM, Mountcastle VB (eds) The mindful brain. MIT Press, Cambridge, pp 51–100;
   Edelman GM (1981) In: Schmidt FO (ed) Organization of the cerebral cortex. MIT Press, Cambridge, pp 535–563
5. Hebb DO (1949) The organization of behaviour. Wiley, New York
6. Silverman DJ, Shaw GL, Pearson JC (1986) Biol Cybern 53 (in press)
7. Wolfram S (1983) Rev Mod Phys 55:601–644; (1984) Nature (London) 311:419–424
8. Proc Conf Chem Based Comput Design (1984) Yates F (ed) UCLA

9. See e.g., the references in the following reviews of neural network models: Harmon LD, Lewis ER (1966) Physiol Rev 46:513–519;
MacGregor RJ, Lewis ER (1977) Neural modeling. Plenum Press, New York;
Amari S, Arbib MA (eds) (1982) Competition and cooperation in neural nets. Springer, Berlin Heidelberg New York
10. Pearson JC et al (1983) Abstr 238.9 for 13th Annu Meet, Soc Neurosci;
Pearson JC (1985) Ph D thesis (unpublished)
11. Shaw GL, Rinaldi PC, Pearson JC (1983) Exp Neurol 79:293–298
12. Morrell F, Hoeppner TJ, Toledo-Morrell L de (1983) Exp Neurol 80:111–146;
Morrell F (1967) In: Quarton GC, Melnechuk T, Schmitt FO (eds) The neurosciences: A study program. Rockefeller Univ Press, New York, pp 452–469
13. Little WA (1974) Math Biosci 19:101–120;
Shaw GL, Vasudevan R (1974) Math Biosci 21:207–218;
Little WA, Shaw GL (1975) Behav Biol 14:115–133;
Little WA, Shaw GL (1978) Math Biosci 39:281–290
14. Shaw GL, Roney KJ (1979) Phys Lett 74A:146–150;
Roney KJ, Shaw GL (1980) Math Biosci 51:25–41;
Shaw GL (1978) Brain Res Bull 3:107–113
15. Shaw GL, Harth E, Scheibel (1982) Exp Neurol 77:324–358;
Shaw GL, Pearson JC (1983) Proc "Orbis Scientiae", Perlmutter A (ed) (in press)
16. Andersen P, Andersson SA (196) Physiological basis of the alpha rhythm. Appleton-Century-Crofts, New York
17. Hubel DH, Wiesel TN (1963) J Neurophysiol 26:1003–1017
18. Merzenich MM et al (1984) J Comp Neurol 224:591–605
19. Verzeano M, Negishi K (1960) J Gen Physiol 43:177–195;
Verzeano M (1980) In: Thompson RF, Hicks LH, Shvyskov VB (eds) Neural mechanisms of goaldirected behavior and learning. Academic, London New York, pp 353–373

Associative Processing in Brain Theory and Artificial Intelligence

A. LANSNER[1]

1 Introduction

The goal of brain theory is to uncover the mechanisms behind biological information processing and intelligence. One hopes eventually to understand the whole range of behavior in animals and in man in terms of the structure and operation of the neuronal circuitry of their brains. To actually create some kind of intelligent device which is the aim artificial intelligence (AI) is clearly a different goal. Yet, it is obvious that these two fields are closely related.

In fact, the early attempts within AI, in the 1940's, 1950's, and 1960's were devoted to the idea of creating machine intelligence by mimicking biological information processing at the level of neuronal circuitry (see e.g. McCulloch and Pitts 1943, Rochester et al. 1956). Essentially, the aim was to construct a device which was analogous to a neuronal network and whose operation was in some sense intelligent. These activities were parallelled with a great interest in theories of brain function and much effort was spent trying to formulate a unifying theory of intelligence. In the beginning, enthusiasm around these ideas was great but as useful results failed to show up, this gradually turned into scepticism. In the later 1960's the time had come for a shift in paradigm. AI entered a new era characterized by a top-down, symbolic-logical approach to intelligence which still to a great extent dominates the field.

Many of the early brain models were based on a close analogy between logical machines and the brain. However, among brain researchers and others seriously concerned with brain function this did arise considerable discomfort. Many of them felt that the analogy between contemporary computers and the brain was misleading and often pushed much too far. The new trends in AI in the 1970's consolidated the separation between the fields and thereafter they have merely coexisted with very little interaction.

The situation is still essentially the same, although there are some encouraging signs of a change. This is not surprising, since it is easy to find arguments to support the view that both AI and brain research would benefit from a closer

[1] Department of Numerical Analysis and Computing Science, The Royal Institute of Technology, Stockholm, Sweden

Brain Theory
Edited by G. Palm and A. Aertsen

cooperation. The last to decades have seen a tremendous development within the field of neuroscience. The body of data produced by experimental brain research is enormous and it is still growing rapidly. Considering the extreme complexity of the system under study it is hard to see how it could be possible to fit all these pieces together without substantially increasing the theoretical efforts.

Furthermore, progress in brain research have had a noticeable impact on AI. An increasing number of people working with AI now feel that it was perhaps not that easy after all to do AI without guidance from biology. The respect for the complexity and ingenuity of biological intelligence is increasing. In fact, some sub-disciplines of AI, e.g. computer vision are clearly moving towards computational structures resembling their biological counterparts (Brown 1984).

Thus, although the ultimate goals of brain research and AI differ, there is a definite convergence in terms of means to reach them. To design nerve net models and study their properties by simulation is becoming an increasingly useful tool in brain theory. However, a successful working model of some significant part of the brain would not only be a major contribution to brain theory but to artificial intelligence as well. Analogously, the AI researcher who manages to design a competitive AI program applying principles similar to those behind biological information processing would certainly deserve appreciation from people working with brain theory.

2 On Associative Information Processing

Obviously, computers are quite good at certain tasks often considered to require intelligence such as e.g. arithmetics, logics, chess etc. On the other hand, problems that the brain solves almost instantly and without any apparent effort e.g. vision and motor control still defeat the most clever AI program running on a powerful computer. One consequence of this is that, paradoxically enough, today a computer may appear more intelligent when compared to a master chess player than to a cat. Many of the problem that are hard for computers seem to involve pattern recognition in one form or the other.

Such observations suggest that biological information processing is not at all or only to a very limited extent based on what can be characterized as digital-logical operations. It seems as if the importance of analytical-logical thinking has been greatly exaggerated in the era of computers. When studying human memory and thought processes more than two-thousand years ago the great Greek philosopher Aristotle formulated the classical laws of association. They describe human thought processes in terms of associations and they are still considered to hold a lot of truth. Actually, humans think logically to a rather limited extent. Even mathematicians have witnessed that a great part of their mathematical thinking does not appear to be of an analytical-logical nature. It is rather experienced as "intuitive", "holistic" or "associative". In particular, original ideas mostly arise in this manner but formal analysis is required when one attempts to prove their correctness.

One of the earliest and more concrete theories of associative brain functions was Hebb's theory of cell assemblies (Hebb 1949). But no doubt, similar thoughts had been expressed even before that (see e.g. Dalenoort 1982). The ideas formulated by Hebb have been elaborated and extended by several researchers. For a review of early theories of associative recall see, for instance, Longuet-Higgins et al. (1970). Today there are several computational models in brain theory, AI and pattern recognition that can be referred to the family of associative models and theories of the brain (Kohonen 1972, 1977, 1984, Wigström 1975, Willwacher 1976, 1982, Anderson et al. 1977, Anderson JA 1983, Braitenberg 1978, Palm 1980, 1981b, 1982, Geman 1981, Lansner 1982, Feldman and Ballard 1982). There is a spectrum from biologically to more technically oriented designs. They are certainly different and have been given different names, but their common characteristic stand out clearly.

These models all recognize the fundamental difference between the brain and conventional computers both at the level of hardware and in terms of operation. They can be fit into a distributed computational structure of neuron analogues operating in a parallel fashion and communicating through connections, i.e. synapse analogues. This network structure is fed with input patterns on which it operates. As we will see, the operations performed are different from those of a logical machine. The system learns by experience and learning rules determine how its structure is altered by ongoing activity, i.e. the information fed into it. This "world knowledge" is stored as associations between objects and events.

The principles of associative pattern processing seems to be extensively used in biological information processing and they have a potential for interesting technical applications. The generality and potential power of "associative processing" will hopefully be even clearer when we in the following describe in some detail an associative network model. Following this we will also discuss further to what extent such a model can serve as a brain model and, finally, we will dwell for a while on the future prospects of associative information processing.

3 An Associative Network Model

Our associative net is composed of a large number of nodes (processors) all operating in parallel and interacting with each other and the outside world through connections. Each node operates as a rather simple threshold element generating its output as a function of the influence received via connections from other nodes in the net. The connections are either reinforcing (excitatory) or inhibitory. Input to the net is provided by driving a set of nodes and thereby activating an input pattern which is assumed to be stable for a sufficiently long time to allow the activity of the net to reach equilibrium.

The nodes are engaged in two fundamentally different kinds of activities. The first one concerns the dynamic flow of activity in a running network. The second very important task relates to the structural changes in the system responsible for learning. This process is guided by certain "learning rules" and it forms the basis for the adaptive (self-organizing) capabilities of the associative net. The learning rules in associative models are generally such that clusters of similar patterns in the input will induce the formation of groups of strongly connected nodes corresponding to Hebbian cell assemblies.

These internal representations are crucial for the operation of associative structures and they have been related to what psychologists refer to as ideas, concepts and Gestalt phenomena. In the literature on associative models they have been given many different names apart from "cell assembly" as suggested by Hebb, e.g. category, neuronal ensemble, stable collection, compacta, template, cluster representative, prototype pattern. Here we prefer to call these internal representations "cell assemblies", "templates", or "concepts" depending on context.

3.1 Associative Operations

An associative net that has been trained with a set of input patterns will be able to perform a couple of powerful and general operations on patterns. These operations can be briefly characterized as follows:

(a) pattern reconstruction and matching
(b) pattern recognition
(c) sequential association

The pattern reconstruction is based on the presence of templates formed during the training session. If a distorted input pattern is fed into the net it is used as a "key" to activate the most similar template, i.e. a matching takes place. This may include completion of the pattern or deactivation of nodes that should not be there.

Recognition of patterns can be accomplished by associating the input pattern to be identified to a pattern representing a symbolic name in another net. Sequential association between concepts is based on the existence of reinforcing connections between them. To learn a sequential association between two concepts the training set has to be made up of pairs of patterns. The first pattern in each pair represents the key concept and the second the associated concept. In this manner associative chains may be stored (see e.g. Willwacher 1982, Lansner 1982).

3.2 Dynamic Flow of Activity

The activity in a running associative net can be described as a succession of active templates, each one initiated, reconstructed, stabilized and finally terminated. Even if a large number of templates are stored in one net, they are only active one at a time. The extent of a template may be somewhat fuzzy and context dependent. Exactly what template that is activated is determined jointly by input to the net and by knowledge stored in its connectivity, i.e. by a best match to current input and by context from previously active templates. It may be noted here, that the reconstruction phase in a running net is related to relaxation methods used in AI especially in computer vision research (Davis and Rosenfeld 1981, Hinton and Sejnowskij 1983).

The positive feedback provided by the connections holding the template together has to be balanced in some way in order to ensure a nice performance and to prevent outbreaks of uncontrolled activity. Various forms of threshold control are used, e.g. inhibitory nodes providing a simple form of lateral inhibition. In fact, the connections are often sufficiently strong to maintain activity in a template even if the driving input pattern is removed. Furthermore, this ongoing activity provides a strong inhibitory surround effect. Generally, the net has to be forced out of this activity once it has been initiated.

Intuitively one feels, that associative recall in such a truly parallel system will be quite fast. Some time is certainly needed to reach steady state after the net has been fed with a new input pattern. But relying on, for instance, the analogy with real neural nets, one simply assumes that this will occur fast enough. Although neurons and synapses are comparatively slow processing elements the nervous system has a reaction time around some hundred milliseconds.

However, this question also have some relevance if one wishes to study these highly parallel systems by means of computer simulation. Since conventional computers are basically sequential machines, simulation programs run very slowly. Thus, a somewhat deeper investigation into the speedup possible to achieve by means of parallel operation may be motivated.

Yet, few theoretical estimations of the speed of recall in associative nets have been done. Since it is not yet known exactly how to design a truly parallel associative net a theoretical treatment of such a system is not feasible. Instead, one has to try to study parallel versions of methods for associative recall currently used in simulations. Lansner and Ekeberg (1984) have considered some such algorithms for a net with all-to-all binary connections. The results indicate that with N physical processors (nodes) associative recall can be performed at least in time $O(log^2 N)$. This suggests that even in systems with many nodes the dynamics will be sufficiently fast.

3.3 Specific Coding and Distributed Representation

Issues of how to represent "world knowledge" in intelligent systems have been subject to intense discussion within the AI society for several years (see e.g. Kobsa 1984). It is also of high relevance to a discussion on associative nets.

The principles of representation applied in our associative network model is very similar to those used in the central nervous system. Here individual sensors often respond selectively in a narrow magnitude interval or to specific forms of stimuli. Rather than being a measure of magnitude as such, the firing frequency of a sensory neuron often signifies a probabilistic measure, i.e. the confidence with which the stimulus to which it is tuned is observed in the input. In this manner, different aspects of a stimulus are mapped onto a spatially organized array of sensors. Only a few will be responding and the rest silent. The difference between two such patterns may be defined as the so called Hamming distance, i.e. the number of nodes that has to be changed to transfer one pattern into the other.

The type of coding discussed above is sometimes in a biological context called "specific" coding. Other names that have been used are "spatial" or "distributed" coding. By AI people it is often referred to as "iconic" or "direct" representation. These representations are rather different from what one is commonly used to in science and engineering, where information dense, scalar and symbolic representations dominate.

One attractive property of the representation proposed here is its homogeneity. Even if two nets process different kinds of information (e.g. images, sound, text, touch) the representations are patterns of active nodes in both nets. This makes the combination of information of different kind and from different sources in a multi-net system quite straightforward.

A general problem of "iconic" representations is their lack of invariance properties. It is obvious that an iconic code in e.g. a system working with vision will be sensitive to translation, rotation and changes in size of objects viewed. Some degree of invariance is certainly desirable but it is not entirely obvious to what extent it should be built into the network. There are different ways to cope with this problem. Invariance with respect to translation can easily be accomplished by using a scanning mechanism that focuses attention on different regions in the visual field. A size invariant iconic representation can be produced by having receptors with radially elongated receptive fields. Furthermore, in an identification task it is possible to connect different patterns to the same label. This reduces the problem of invariance to a matter of memory requirements and degree of generalization in the learning process.

3.4 Storage Capacity

Obviously, the storage capacity of associative nets is a crucial question both for their plausibility as models of the brain and for their practical usefulness. An efficient memory utilization requires that the amount of information possible to retrieve is linearly related to the number of physical memory elements used. Thus, the redundancy needed for e.g. pattern completion may not grow faster than linearly with the number of memory elements.

In fact, a limited storage capacity has sometimes been claimed to be a major draw-back of associative memories. This may be due to a mis-interpretation of considerations relating to linear models. In such systems a total reconstructive power can be achieved if the patterns stored are orthogonal. Then, however, it is only possible to store as many patterns as there are nodes in the net whereas the physical memory of the system depends on the square of the number of nodes. However, in this case, there is no simple relation between the number of patterns and the amount of information stored. Therefore, this result may not be interpreted simply as a low storage capacity in the above given sense.

On the other hand, some information theoretical calculations based on a non-linear associative memory model with binary connections have also been carried out (Palm 1980). They show, that the information stored grows linearly with the number of memory elements (connections) and may asymptotically reach 69% of the theoretical limit. The number of patterns one can store in this case grows almost as the square of the number of nodes in the system. Furthermore, an efficient memory utilization is possible already in reasonably small systems. For this to hold the number of nodes activated by a pattern should be logarithmically related to the number of nodes in the net.

Lansner and Ekeberg (1984) recently studied the trade-off between reconstructive power and reliability of recall in a similar system. Their results show

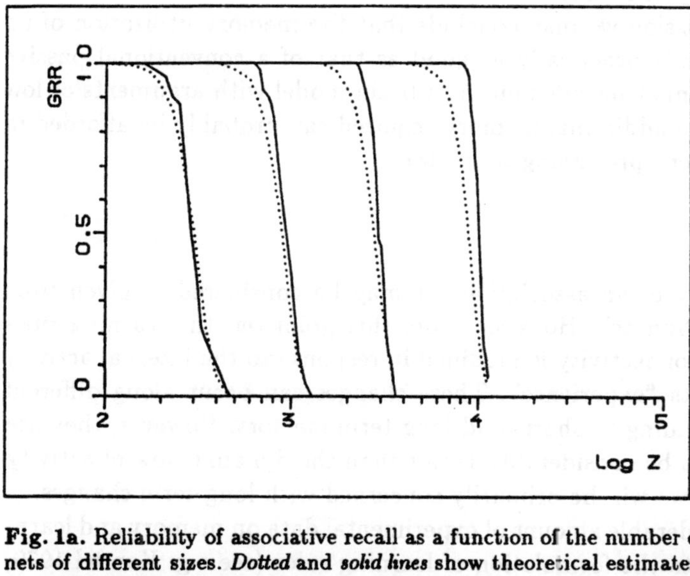

Fig. 1a. Reliability of associative recall as a function of the number of patterns stored (Z) in nets of different sizes. *Dotted* and *solid lines* show theoretical estimates and simulation results respectively. From *left* to *right* net size was 250, 500, 1000 and 2000 nodes. Global recall reliability (GRR) is the fraction of the stored patterns that could be correctly recalled. The patterns stored contained between 14 and 18 nodes. (From Lansner and Ekeberg 1984)

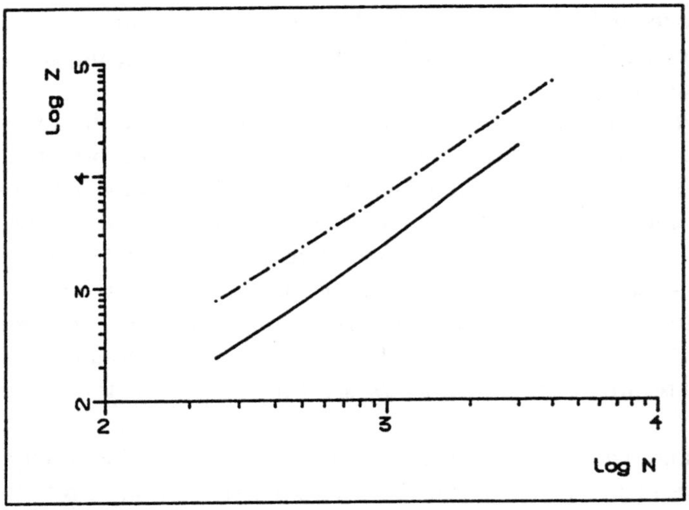

Fig. 1b. Growth of storage capacity (Z = number of patterns stored) with net size at a recall reliability of 90%. *Solid lines* shows simulation results for net sizes 250, 500, 1000, 2000 and 3000 nodes. *Dash-dotted line* shows information theoretical limit without regard to recall reliability

that a high storage capacity is possible without sacrificing recall reliability. Figure 1a shows that the associative recall process is perfect over a wide range and displays a clear saturation effect. If the number of patterns stored is increased beyond a certain point the recall reliability drops dramatically. Figure 1b shows how the number of patterns stored at a certain reliability of recall (90%) grows with system size.

From this discussion we may conclude that the memory utilization of an associative structure is practically as good as that of a conventional passive memory. Thus, it cannot be ruled out as a brain model with arguments of low storage capacity. The additional memory required can probably be afforded in many technical pattern processing tasks, too.

3.5 Learning Rules

Part of the structure of an associative net may be considered as given from the start, i.e. it is "innate". However, from this point on, the learning rules determine how the connectivity is modified in response to the internal activity of the system, i.e. its "experience". These changes can occur along different time-scales corresponding to short- and long-term memory. However, they are generally assumed to be considerably slower than the dynamic flow of activity in the system. Here we will be primarily concerned with long-term changes.

There is a considerable amount of experimental data on memory and learning that is relevant to the formulation of learning rules (see e.g. Kandel 1979, Kupfermann 1981). Unfortunately, it appears as if this knowledge is not yet complete and detailed enough to provide a firm basis for the formulation of quantitative models. Thus, since biology cannot yet tell us exactly how to design appropriate learning rules, we will have to rely on performance type of criteria for the time being. Reasonable candidates for learning rules have to be analyzed theoretically and tried out by means of simulations.

When discussing learning rules it is convenient to distinguish between changes related to template formation, development of higher order nodes, and learning of sequential associations respectivley. Here we will concentrate on the two first of these and treat them separately below. Despite this, there is a close relation between these different processes and it may well be that the learning rules in some cases eventually will turn out to be identical.

3.6 Template Formation

In their simplest form associative nets have binary connections and the nodes are allowed to connect in an all-to-all manner ("switching matrix"). In this case, the learning rules cannot be very sophisticated. These models are commonly employed in studies of storage and retrieval such as those mentioned in the section on storage capacity above. However, the capabilities demonstrated by these simple systems are not sufficient in more realistic pattern processing tasks. The main problem seems to be, that the real world does not provide the templates required by this scheme. Instead, these have to be extracted from "raw data" by means of a statistical clustering process that separates significant and reproducible properties of the world from mere coincidences. Clearly, there is a qualitative difference between a system capable of storing, for instance, ten thousand pattern instances supplied by some external operator and one that extracts the same number of different templates from the environment in a non-supervised learning process.

These highly attractive "template formation" capabilities are commonly attributed to associative structures. However, the question of what kind of learning rules that realize such functions has not yet been answered satisfactorily. A number of schemes have been proposed, most of them based on the principle of two-conditional facilitation put forward by Hebb (Hebb 1949). He suggested, that simultaneous activity, or nearly so, in two neurons was a necessary condition for the formation and subsequent enhancement of excitatory synapses between them. In fact, the existence of such Hebbian synapses in the nervous system has recently gained some experimental support (Levy and Steward 1979). Several variations and quantitative formulations of Hebb's original hypotheses have been forwarded.

For instance, the idea that the strength of a connection between two nodes in an associative structure should in some way reflect the correlation of their activities has been put forward and theoretically elaborated by several authors (Uttley 1958, Marr 1970, Kohonen 1972, Geman 1981). This kind of learning rule incorporates both potentiation of frequently used connections and weakening and elimination of unused ones (forgetting). Furthermore, the resulting steady state connectivity will reflect the probabilities and hierarchical character of clusters in the real world.

The consequence of applying this kind of learning rules on a set of reasonably realistic input patterns can be theoretically predicted for a linear system. Analysis and simulations of such models has been carried out (Kohonen et al. 1977, Anderson et al. 1977, Anderson and Mozer 1981, Geman 1981). However, since the model discussed here is essentially nonlinear and use a different type of representation these results are not entirely relevant to us.

The learning rules underlying template formation in our model are of a similar kind and works as follows: Each time two nodes are activated by the same input pattern the connection between them is increased by a fixed value. Furthermore, the strength of all the connections originating from a particular node, decays exponentially. The half-time of this decay is inversely proportional to the sum of the strengths of all the connections from that node. This means that the sum of the connections originating from a node will tend to approach the same value for all nodes in the net. The strength of a connection from node i to j will approximate the conditional probability of j being active given that i is active, i.e. $p(j|i)$.

We have performed some simulation experiments demonstrating a primitive form of template formation in such a system. Input patterns were simple hand-drawings fed from a tablet coded into the net which was a 21×21 matrix of nodes. All the figures were of about the same size and position. There were

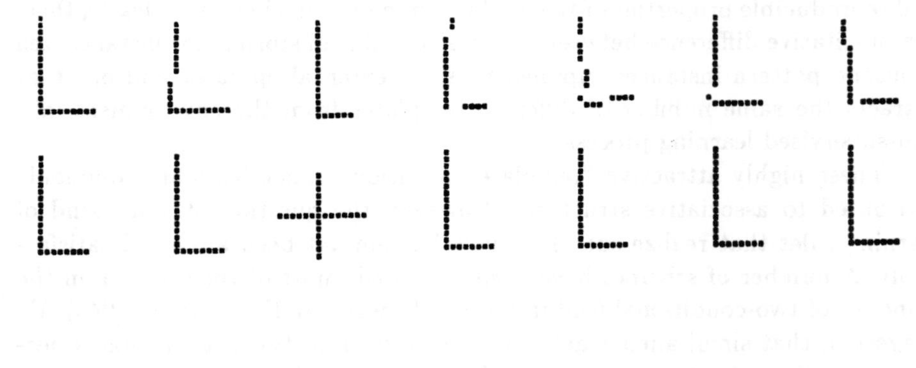

Fig. 2. Template formation in a single layer associative net. Twenty samples of L and *cross* were used both as training and test set. The *first* and *third rows* show a subset of these input patterns. The *second* and *forth* rows show results from the test session for these patterns. It can be seen that one L- and two *cross*-templates have formed. In 38/40 cases a correct template was activated

two different types of figures and the training set were composed of twenty instances of each type.

After a training session followed a test session in which the same set of figures were once more used as input. Each input pattern was iterated to stability. In the majority of cases, this process converged to a stable configuration representing an idealized version of the corresponding type of figure (Fig. 2). This kind of reconstructive matching was not limited only to figures fed into the net during the training session. Several other new instances of the same type of figures as those in the training set also activated the correct template.

3.7 Higher Order Nodes

The notion of complex cells has been well known to brain research for many years. Cells with response properties ranging from simple to complex and hypercomplex have been observed in many parts of the central nervous system, e.g. in the mammalian visual system (see e.g. Hubel and Wiesel 1979). Complex cells are assumed to be driven by the simpler ones in a hierarchical manner with the receptor cells at the bottom. This organization is known to be dependent on exposure to stimuli, especially during certain "critial periods" in young animals (Hubel and Wiesel 1965). How far this neuronal specificity actually reaches, i.e. whether cells with extreme complexity and selectivity (sometimes called grandmother cells) exist or not is a classical question which has not yet been answered.

The mechanisms behind the development of neuronal specificity have been extensively discussed and modelled (see e.g. von der Malsburg 1973, Perez et al. 1975). These studies relate mainly to the visual cortex where experimental data seem to be most abundant. Computational models of this kind of vertical organization have also been a recurrent property of associative type of models. For instance, early theories of perceptrons included "association units" of increasing complexity (Rosenblatt 1961). More recently such principles have been discussed e.g. by Geman (1981), Feldman and Ballard (1982), and Fukushima (1984).

Here the complex nodes are referred to as "higher order nodes". The order of a node is defined such that the simplest nodes are of order one, and the order of other nodes is equal to the number of first order nodes on which their response depends. This forms the basis for talking about different levels and layers in the net with the primary layer containing first order nodes, the secondary layer second order nodes etc.

Higher order nodes represent higher order correlations in the input patterns. Consider a number of classes of patterns each represented by a template in the net. In the lower layers many templates may overlap considerably but in higher layers with more selectively responding nodes overlap will generally be smaller (Fig. 3). This decreases cross-talk between templates and allows the associative operations to work more reliably than if these layers were absent. A recent study on the storage capacity and reconstructive power in associa-

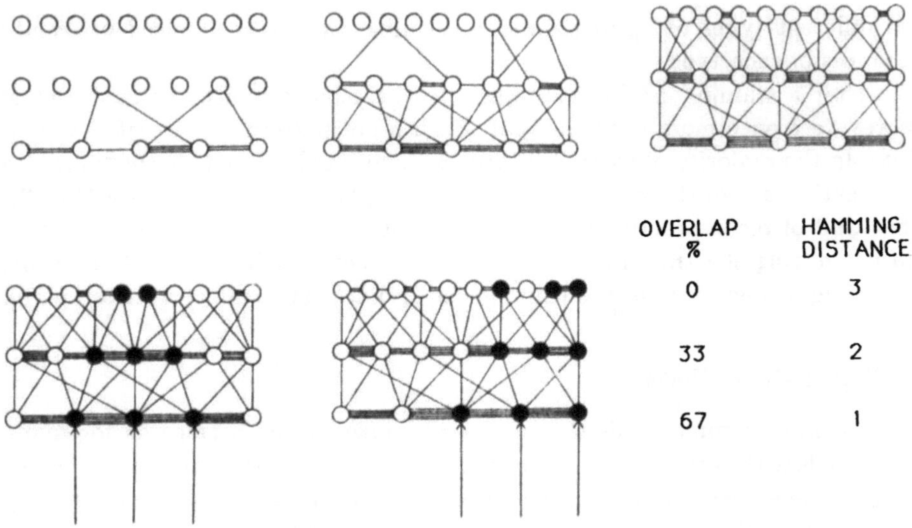

Fig. 3. *Upper half* schematic illustration of the development of a small multi-layer associative net with learning (from *left* to *right*). The structure grows vertically by formation of nodes of higher order (<= 3) and horizontally by an increasing density of connections within each layer. *Lower half* distributed representation of two different but overlapping input patterns. Notice how the overlap decreases and Hamming distance increases towards the upper layers

tive memories storing third order correlations by Shiozaki (1984) illustrates this effect. Higher order nodes also provide a more specific signature useful for pattern recognition tasks.

It is clear that the potential number of higher order nodes is tremendous if no restrictions on their creation are imposed. The overall goal of the learning rules should be to recruit nodes selectively such that an efficient composition of nodes of different order is obtained within the limits of the physical memory available. In our simulation program, connections related to template formation may only connect nodes in the same layer. Higher order nodes are created when the horizontal connections between two lower order nodes becomes sufficiently strong. The lower order nodes also have to be close enough to be allowed to generate a higher order node and no nodes with order above some limit, e.g. five, are allowed. Using these rules it is possible to modulate the vertical structure generated from a certain set of training patterns within wide margins by varying the values of a few parameters. The principal structure of the net produced is outlined in Fig. 3.

Some preliminary simulations have been carried out using this multi-layer associative net. Figure 4 shows the higher order nodes generated in a simulation of the same task as described above for template formation although nodes up to the order of five were allowed to form. A pure pattern recognition task has also been studied. Preliminary results show that this system can discriminate between at least thirty different types of figures. As expected, rotation and size invariance is poor in the recognition.

Fig. 4. The response characteristics of nodes in different layers of an associative net. A drawing of a square was fed into the net and the figure shows the driving stimulus of a small subset of the nodes activated in layers 1 (*bottom*) to 5 (*top*)

4 Associative Structures as Brain Models

Clearly, brain theory has not yet reached the point where one has agreed upon a unified and generally accepted model of the brain that is reasonably complete and capable of quantitative and experimentally testable predictions. On the contrary there are still several competing models that propose radically different mechanisms to explain the same phenomena. Furthermore, all of them are incomplete and more or less of a qualitative nature.

Even the class of models based on associative processing that we concern ourselves with here is highly heterogeneous. Several possible designs have been put forward and many of the central issues discussed in the previous section are still subject to investigations. There is, however, a common general framework accepted by those working with this particular set of models. The description in the previous section showed that the development of the theory of associative structures is intimately related to progress in brain research. Here we will discuss some more concrete motives for their use as models of the brain. To begin with the structural parallels will be considered and thereafter we treat those of a more functional character.

4.1 Structural Aspects

The anatomical analogies between the brain and the associative structures are rather obvious. The neurons and synapses of the brain have their counterparts in nodes and connections in the models. Certainly, the biological building blocks are much more complex than their analogues. However, it is reasonable to assume that those properties of the neuronal network that are most essential for the information processing carried out are also represented in the model.

A more specific and highly significant parallel is the mechanisms behind memory and learning. In associative nets, it depends on creation of connections and modification of their strength according to the learning rules. This corresponds well to the fact that, today, there is almost unanimous agreement among workers in the field of neuroscience that the functional plasticity of synapses is a primary physiological correlate to memory and learning (see e.g. Kandel 1979, Kupfermann 1981).

Above we discussed two rather different kinds of structural changes, i.e. those related to formation of higher order nodes on one hand and to template formation on the other. A lot of work in neurobiology is related to the mechanisms behind development of neuronal specificity which would correspond to the formation of higher order nodes. In the brain as in our model, these processes are affected by early sensory experience. Furthermore, in the mammalian cerebral cortex a long-range (centimeter) myelinated cortico-cortico and a short-range (millimeter) non-myelinated fiber system have been described. These have been called the *A*- and *B*-system respectively by Braitenberg (1978). One may assume that the long-range *A*-system is required to connect the different parts of a cell assembly whereas the *B*-system primarily supports the development of complex cells driven by nearby simple cells.

There is also one notable difference between the biological system and the model for formation of higher order nodes proposed here. Real neurons often display a progression towards more selectivity in some respects and more generality in others at the same time (e.g. shape vs. position of driving stimulus). In the model discussed the higher order nodes just become more and more selective. There are, however, other similar models of these phenomena that are more realistic in this respect.

4.2 Functional Aspects

The operations characteristic of associative structures are closely related to many important aspects of brain function. Learning, concept formation, pattern reconstruction and matching together with sequential association are all important features of the working brain. Furthermore, distributed and redundant storage of memory traces is a well-known property of the brain demonstrated, for instance, by Lashleys classical experiments. Among other things, it results in a good robustness against damage which is also characteristic of artificial associative structures (Palm 1981a).

According to the theory of associative nets, the major vehicle of thought processes is the concept represented by a cell assembly. Clearly, the case for using these structures as brain models would be considerably stronger if one could find concrete anatomical or functional evidence for the existence of cell assemblies in the brain. However, the observation of a cell assembly by means of neuro-physiological recordings is probably a very unlikely event. The synapses holding a cell assembly together need not be very strong and the membership of neurons in cell assemblies are likely to be fuzzy and context dependent. Probably multi-neuron recordings to an extent which is not yet technically feasible would be required.

One kind of quantitative measurements that might reveal something about cell assembly dynamics is electro-physiological recordings of stimulus related mass-potentials, e.g. evoked potentials (EP). It seems reasonable to assume, that the electrical activity in the neural tissue related to the "ignition" of a cell assembly (concept) is sufficiently strong to show up in such registrations. Actually, in a recent study (Glass and Hall 1981) it was suggested that slow waves recorded from cat visual cortex actually reflect ongoing pattern processing. Responses occurring between 50 and 300 ms after stimulus onset were affected by the type of pattern presented. Other studies have demonstrated the influence of stimulus significance on these potentials (Cooper et al. 1977, Grastyan et al. 1978).

Some results relevant to the dynamics of cell assembly activity have also been produced in computer simulations of a reasonably realistic model of a small piece of mammalian cortex (Lansner 1982). They showed that the time-course of ignition of a cell assembly would be in the order of ten to a hundred milliseconds and that activity is terminated by means of accumulated slow hyperpolarization or synaptic fatigue after some further two hundred milliseconds. Another interesting observation was that model neurons with parameter values corresponding to motoneurons and to pyramidal cells behaved differently. Pyramidal cells which have a much weaker post-spike hyperpolarization than motoneurons seemed to be better suited for participating in cell assembly related activity.

A critical question when discussing associative structures as brain models relates to if and how higher brain functions in humans such as e.g. language and logical reasoning can be fit into this scheme. The influence of computer science on brain theory and psychology is powerful and still many models of the brain bear strong resemblances with computer organization. This tendency is particularly accentuated when higher functions of the human brain are considered. Whereas associative pattern processing is by many quite readily accepted as underlying low-level processing (e.g. early vision), its heuristic and probabilistic character make it less plausible as a model for these higher functions.

However, the brain-computer analogy is probably misleading in this case, too. As already pointed out, the intuitive, associative components of intelligent behavior are highly significant in animal as well as in man. Thus, it has been claimed that associative processing is the basis even of thought processes in humans (Anderson JA 1983, Kohonen 1984). However, there is so far very few hard facts to support this notion. An interesting indication, though, is the existence of some psychological models of thought processes and memory in humans that approaches the associative models. They have been proposed to explain observations from psychological experiments such as e.g. interference results in memory, judgment of associative relatedness, impact of extensive practice on memory, effects of reconstructive recall, and sentence comprehension (Anderson JR 1983, Gigley 1982, Small et al. 1982).

It seems reasonable to conclude that the associative theories and models still retain and actually have further consolidated their position as the most promising potential brain models, so far. They represent an attempt to bridge the wide gap between our extensive knowledge of the micro-structure of the brain on one hand and the phenomenological manifestations of brain function on the other. Support for their plausibility is slowly accumulating and at some points the analogies and qualitative predictions are comparatively concrete. However, much work remains, since there is still a considerable distance between these theories and the reality of experimental brain research.

5 The Future of Associative Processing

Undoubtedly, the associative operations we have discussed here would be very attractive in many artificial intelligence and pattern processing tasks. However, our knowledge concerning associative processing is still not sufficient to allow for any full-scale applications. Some central research objectives for the near future include the study and design of adequate learning rules, mechanisms for short-term memory, goal-directed operation and motivation together with ways of generating and controlling motor output.

Furthermore, to formulate models of higher functions related to language and reasoning based on associative processing is another important goal. When discussing brain modelling above we suggested that this scheme can presumably support such functions but that this remains to be shown. Moreover, even if high-level functions can be accomplished in this manner it may still not be the most efficient solution. Perhaps hybrid systems equipped both with conventional arithmetical and formal-logical capabilities and associative mechanisms of the kind discussed here will turn out to be the most competitive in the future.

A bottleneck in both basic research on associative structures and in applications is the current lack of suitable parallel hardware. These algorithms run notoriously slow on sequential computers and a considerable speed-up could be expected from massively parallel hardware. Thus, development of parallel machines is a critical step in making associative processing a realistic alternative to more conventional forms of information processing. The highly distributed structure of an associative net, the asynchronous operation of its nodes, the local learning rules and the tolerance against hardware errors are all attractive properties of algorithms that one wants to transfer to parallel digital-logical hardware. In fact, at least one research group have already designed a small parallel associative computer based on a set of conventional microprocessors communicating via a common bus (Palm and Bonhoeffer 1985).

Thus, it may only be a matter of time until reasonably powerful purely associative computers exist. They will make up a new generation of machines with essentially complementary properties compared to computers as we know them today. Instead of being programmed in a conventional manner, they will be capable of learning from experience and by means of training. What one computer has learned may be transferred "in extenso" to its successors. If and when this kind of computers come into existence they will certainly open up new perspectives in information processing.

References

Anderson JA (1983) Cognitive and psychological computation with neural models. IEEE Trans Syst Man Cybern SMC-13:799–815

Anderson JA, Mozer MC (1981) Categorization and selective neurons. In: Hinton GE, Anderson JA (eds) Parallel models of associative memory. Lawrence Erlbaum Assoc, pp 213–236

Anderson JA, Silverstein JW, Ritz SA, Jones RS (1977) Distinctive features, categorical perception, and probability learning: Some applications of a neural model. Psychol Rev 84:413–451 (1977)

Anderson JR (1983) A spreading activation theory of memory. J Verbal Learning Verbal Behav 22:261–295

Braitenberg V (1978) Cell assemblies in the cerebral cortex. In: Heim R, Palm G (eds) Theoretical approaches to complex systems. Springer, Berlin Heidelberg New York, pp 171–188

Brown CM (1984) Computer vision and natural constraints. Science 224:1299–1305

Copper R, McCallum WC, Newton P, Papakostopoulus D, Pocock PV, Warren WJ (1977) Cortical potentials associated with the detection of visual events. Science 196:74–77

Dalenoort GJ (1982) In search of the conditions for the genesis of cell assemblies: A study of self-organization. J Social Biol Struct 5:161–187

Davis LS, Rosenfeld A (1981) Cooperating processes for low-level vision. Artif Intell 17:245–263

Feldman JA, Ballard DH (1982) Connectionist models and their properties. Cognitive sci 6:205–254

Fukushima K (1984) A hierarchical neural network model for associative memory. Biol Cybern 50(2):105–113

Geman S (1981) Notes on a self-organizing machine. In: Hinton GE, Anderson JA (eds) Parallel models of associative memory. Lawrence Erlbaum Assoc, pp 237–263

Gigley HM (1982) A computational neurolinguistic approach to processing models of sentence comprehension. COINS Tech Rep 82–9, Univ Massachusetts, Amherst

Glass JD, Hall RW (1981) Pattern processing and slow-wave activity in visual cortex of cat. Electroenceph Clin Neurophysiol 52:116–126

Grastyan E, John ER, Bartlett F (1978) Evoked response correlate of symbol and significance. Science 201:168–171

Hebb DO (1949) The organization of behavior. Wiley, New York

Hinton GE, Sejnowski TJ (1983) Optimal perceptual inference. Proc IEEE Conf Comput Vision Pattern Recognition (CVPR), Washington DC

Hubel DH, Wiesel TN (1965) Binocular interactions in striate cortex of kittens reared with artificial squint. J Neurophysiol 28:1041–1059

Hubel DH, Wiesel TN (1979) Brain mechanisms of vision. Sci Am 24(3):150–162

Kandel ER (1979) Cellular insights into behaviour and learning. Harvey Lect 73:19–92

Kobsa A (1984) Knowledge representation: A survey of its mechanisms, a sketch of its semantics. Cybern Syst 15:41–89

Kohonen T (1972) Correlation matrix memories. IEEE Trans Comput C-21:353–359

Kohonen T (1977) Associative memory – a system theoretic approach. Springer, Berlin Heidelberg New York

Kohonen T (1984) Self-organization and associative memory. Springer, Berlin Heidelberg New York

Kohonen T, Lehtiö P, Oja E, Kortekangas A, Mäkisara K (1977) Demonstration of pattern processing properties of optimal associative mappings. Proc Int Conf Cybern, Soc Washington CD, pp 581–585

Kupfermann I (1981) Learning. In: Kandel ER, Schwartz JH (eds) Principles of neural science. Elsevier/North-Holland, Amsterdam New York

Lansner A (1982) Information processing in a network of model neurons. A computer simulation study. Rep TRITA-NA-8211, Dep Numerical Anal Comput Sci. R Inst Technol, Stockholm

Lansner A, Ekeberg Ö (1984) Reliability and speed of recall in an associative network. Rep TRITA-NA-8408, Dep Numerical Anal Comput Sci. R Inst Technol, Stockholm

Levy WB, Steward O (1979) Synapses as associative memory elements in the hippocampal formation. Brain Res 175:233–245

Longuet-Higgins HC, Willshaw DJ, Buneman OP (1970) Theories of associative recall. Q Rev Biophys 3(2):223–244

Malsburg C von der (1973) Self-organization of orientation selective cells in the striate cortex. Kybernetik 14:85–100

Marr DO (1970) A theory for cerebral cortex. Proc R Soc London Ser B176:161–234

McCulloch WS, Pitts WH (1943) A logical calculus of the ideas immanent in nervous activity. Bull Math Biophys 5:115–133

Palm G (1980) On associative memory. Biol Cybern 36:19–31

Palm G (1981a) On the storage capacity of an associative memory with randomly distributed storage elements. Biol Cybern 39:125–127

Palm G (1981b) Towards a theory of cell assemblies. Biol Cybern 39:181–194

Palm G (1982) Neural assemblies. An alternative approach to artificial intelligence. Springer, Berlin Heidelberg New York

Palm G, Bonhoeffer T (1985) Parallel processing for associative and neuronal networks. Biol Cybern 51:201–204

Perez R, Glass L, Schlaer R (1975) Development of specificity in the cat visual cortex. J Math Biol 1:275–288

Rochester N, Holland JH, Haibt LH, Duda WL (1956) Tests on a cell assembly theory of the action of the brain, using a large digital computer. IRE Trans Inf Theory IT-2:80–93

Rosenblatt F (1961) Principles of neurodynamics: perceptrons and the theory of brain mechanisms. Spartan

Shiozaki A (1984) Recollection ability of three-dimensional correlation matrix associative memory. Biol Cybern 50(5):337–342

Small SL, Cottrell GW, Shastri L (1982) Toward connectionist parsing. Proc Net Conf AAAI, Pittsburgh

Uttley AM (1958) Conditioned probability computing in a nervous system. In: Mechanization of thought processes. HMSO, London

Wigström H (1975) Associative recall and formation of stable modes of activity in neuronal network models. J Neurosci Res 1:287–313

Willwacher G (1976) Fähigkeiten eines assoziativen Speichersystems im Vergleich zur Gehirnfunktion. Biol Cybern 24:181–198

Willwacher G (1982) Storage of a temporal sequence of patterns in a network. Biol Cybern 43:115–126

Associative Networks and Cell Assemblies

G. PALM[1]

1 Introduction

Since the time of McCulloch and Pitts' Theory (1943) there have been many attempts to model the flow of activity in neural networks. It is possible to simulate neural networks (of rather small size) on a computer, relying on quite reasonable – more or less simplified – assumptions on the dynamic behavior of single neurons. One problem is the arbitrariness of the design of the network (i.e. the connectivity matrix). Here many investigations have studied random connectivity (e.g. Anninos et al. 1970, Griffith 1971, Amari 1974, Dammasch and Wagner 1984) or connectivity that itself changes subject to certain rules (for an overview see Palm 1982).

Such simulations may give an impression of the flow of activation through real neural networks, but they reveal a much more severe problem: How should we interpret what we observe in this flow of activation (e.g. as learning or as problem solving)?

In many cases somewhat naive interpretations of the observed network behavior have been given, and these have been duly criticized. Obviously what is needed for the interpretation of such simulations is the relation between the observed activity patterns in the brain and the events in the outside world – either through sensory inputs to the simulated "brain" or through motor outputs from it.

This relation is studied experimentally in the more "peripheral" areas of the brain (for example by Hubel and Wiesel), but the relation of these areas to the classical association areas is still unclear. This problem of the interpretation of the flow of activity in real or simulated networks of neurons, naturally leads to the question of the representation of information or knowledge about the outside world inside the brain. Once this question is answered, one can interprete patterns of activation in the neural network as representing something in the outside world and in this way relate the observed network behavior to operations on such representations that might correspond to learning, thinking or the like.

[1] Max-Planck-Institut für Biologische Kybernetik, Spemannstraße 38, 7400 Tübingen, FRG

Brain Theory
Edited by G. Palm and A. Aertsen

But is this idea of representation of information or even knowledge in a network not too far fetched? Certainly the knowledge of making coffee is somehow built into a coffee-machine, but would we say that it is "represented" there?

In the context of robots, computerized systems and their design one did not talk about the representation of knowledge until very recently. Only for very complex multi-purpose computer programs does the representation of knowledge in them or for them become an issue.

I think this is due to the fact that such a multi-purpose program needs a memory to store the information necessary for its adaptation to one of its many possible purposes or options. This information has to be provided by the user of the system, and if it is a lot, the problem of representing this information inside the program becomes important.

Taking this analogy to animals, I would argue that only for animals that are capable of a highly flexible, non-stereotypic behavior does the representation of information or knowledge inside their nervous system become a real issue. In other words: speaking of representation of knowledge only makes sense for systems or animals that are capable of flexible behavior and thus need a memory, and it should essentially refer to a *central representation of information in the memory*. The representation should be central in the sense that it is action oriented, but still contains the relevant sensory input data. This view also implies that one should not consider a representation simply as an image. Certainly an image can be one kind of representation, which is very close to the sensory input, but there are much more general representational schemes; for example the representation of knowledge in complex computer programs is mostly in terms of relational structures (e.g. Wilson 1980), and the entities that are possibly represented in our memory may be of quite different kinds. They may be images of things or of situations, but they may also be more general concepts or types of situations that cannot be conceived of as images.

One important issue concerning the representation of information in the human memory has always been its *distributed nature*. This has motivated the investigation of *associative memories*, which are technical devices that store information in a distributed way. Moreover, it has often been pointed out that such associative memories can be realized by means of neural networks in a very natural way, if one only assumes that the synapses in these neuronal networks are modifiable (Palm 1982). The issue of the representation of information in such an associative network has also been considered – motivated by Hebb's concept of *cell assemblies*, i.e. by the idea that concepts are not represented in the brain's memory by single neurons, but by interconnected sets of neurons, the cell assemblies.

All this is well known and many simulations have successfully demonstrated the working of associative memories, cell assemblies or neural networks with modifiable connections (for an overview see Palm 1982), all based on very similar equations. For further progress, I think that quantitative evaluation criteria for the comparison of the various similar models are needed. Such a quantitative comparison can only be carried out on a certain level of abstraction, focusing on the common structural principles, (e.g. concerning the formation of assemblies, or their self-completing properties) and disregarding the actual representations that are used in different models (which is still a very important issue concerning the qualitative features of the models).

In this paper I want to define two basic tasks that an associative memory should fulfill, both of which should make sense from an introspective viewpoint and from a purely technical one as well: *pattern mapping* and *pattern completion.* I shall then describe the basic mathematical structure upon which the different approaches have converged, and which can, indeed, be used in a straightforward way to carry out the two tasks of pattern mapping and pattern completion. This structure has appeared in different areas such as pattern recognition and neural network modelling under many different names like "Lernmatrix" (Steinbuch 1961), "conditional probability machine" (Uttley 1956), "cognitron" (Fukushima 1975), "associative network" (Kohonen 1977, Lansner 1983), or as a model for certain parts of the brain like the cerebellum (Marr 1969), the hippocampus (Marr 1971), the neocortex (Marr 1970, Anderson 1972) or the visual cortex (van der Malsburg 1973, Nass and Cooper 1975).

I shall call it the *associative matrix* or the *associative network.* The connection between networks and matrices is rather close: a network of a finite number of elements $e_1, \ldots, e_n$ can always be uniquely specified by its connectivity matrix $C = (c_{ij})$, where the number c_{ij} denotes the strength and sign of the connection from element e_i to element e_j. $(i, j \in \{1, \ldots, n\})$.

It is probably the following fundamental idea that has led different scientists from different fields to consider associative networks: *correlations between the occurrence of "items" or "events" should be stored in terms of connections between units "representing" these items.*

In the resulting network of connections highly connected sets of units would appear (the so-called *assemblies*) and these would then represent combinations of highly correlated items, which may quite rightly be called *concepts* (remembering the original sense of the word).

It turns out that this idea does work quite well, and this to some degree independently of the particular nature of the "representation", i.e. the particular correspondence between events in the outside world and the activation of units in the memory network.

This problem of choosing a reasonable correspondence is, of course, a very important one (I call it the problem of input – output coding), but its analysis can be separated from the analysis of the memory itself, which is the main subject of this paper. As a basis for this analysis and for the comparison of different memory-organizations, the rather new concept of the operational information storage capacity of a memory is introduced and discussed in the last part of the paper.

2 The Task of a Memory

For a general discussion and comparison of various memories, independent of particular contexts and their representation in the memory, we need as a starting point a definition of the task of a memory on the right level of abstraction. The task has to be sufficiently general and universal, so that the memory can be used directly for any concrete information storage problem and yet it has to contain some specification of what is stored and how it is retrieved.

2.1 Pattern Mapping

Here the memory has to store a set S of pairs (x, y) of patterns x and y in such a way that in the recall the memory responds with the output y to the input x, for every pair (x, y) in S.

The patterns x and y may be sequences of symbols (from an alphabet), and in this case it would be convenient if the memory would not only respond with y to x, but also with something similar to y (in the Hamming-distance sense) to something similar to x.

Some data about the problem should be given in advance, I shall refer to them as to the *format* of a particular task. It consists of 3 numbers: the number of all *possible input* patterns Pi, the number of all *possible output* patterns Po and the *actual* number A of input-output pairs contained in the set S.

Let us briefly consider the usual way of storing the information contained in such a specific set S of the format (Pi, Po, A). We could simply write each of the pairs (x^i, y^i) into one "line" of an appropriate storage medium. This storage medium would optimally be organized into A such lines, each of which has the possibility of storing $\log_2 Pi + \log_2 Po$ bits of information (since each bit can inform us about a choice between two possibilities, n bits can specify one among 2^n possibilities, and the number of bits is thus always the "dual" logarithm (to the base 2) of the number of possibilities). When we then want to perform the mapping, we have to search through the stored x^i to see whether the given x is on the list, and if it is, we can produce the corresponding y^i from the same line. This search can be simplified by a proper (e.g., alphabetic) ordering of the x^i as in a telephone-directory. In a sense the $\log_2 Pi$ bits specifying the (known) x^i are wasted in this procedure. It effectively stores only $A \log_2 Po$ bits, but needs room for $A \log_2 Po + A \log_2 Pi$ bits.

There is one interesting way of avoiding this search altogether: We could use a storage medium with Pi lines and a fixed enumeration of the Pi different possible inputs, which can be computed from a simple formula. For example, if the possible inputs are strings $(a_1, \ldots, a_6)$ of length 6 from a 40-letter alphabet, we may as well assume that the a_i are numbers from 1 to 40. Then we can compute the number

$$n(a_1, \ldots, a_6) = a_1 + 40\{a_2 + 40[a_3 + 40(a_4 + 40(a_5 + 40a_6))]\}.$$

This formula can be computed by a programmed calculator in almost no time and a similar machine could automatically point to the corresponding line of our storage medium. With the aid of this machinery we could for a given pair (x^i, y^i) compute the number $n(x^i)$ and store y^i in the corresponding line of the storage medium, from which it can be retrieved (given x^i) by the same procedure. The number $n(x^i)$ may be called the *address* of y^i. A machine that has a storage medium and, given an address, responds with the contents of the corresponding line of the store, is called a "random access memory".

This procedure really needs no searching: it corresponds to a telephone directory containing literally *every* alphabetically possible name; for such a telephone book, one could find a formula to compute for any given name the page on which the name can be found and another formula for the line on that page.

Of course, this procedure of using a storage medium with Pi lines and the possibility of storing $\log_2 Po$ bits in every line is a waste of storage space, if Pi is much larger than A (which is the case in most common applications like the telephone directory). It effectively stores $A \log_2 Po$ bits but needs a storage space for $Pi \log_2 Po$ bits.

2.2 Pattern Completion

Here the memory has to store a set S of patterns x in such a way, that in the recall it responds with the output x to any input-pattern x' that contains a sufficiently large "part" of the pattern x, or that is sufficiently close to x in the Hamming distance, when x is a sequence of symbols.

Let me further specify the pattern completion task. A pattern x shall be a sequence of k symbols or "items", $x = (x_1, \ldots, x_k)$, and for any particular pattern completion task the following numbers should be known (or estimated) in advance: the number P of all *possible* patterns x, the number k of items in a pattern, the numbers P_j of all *possible* j-th items x_j $(j = 1, \ldots, k)$ and the *actual* number A of patterns contained in the set S. These $k + 3$ numbers $P, k, P_1, \ldots, P_k, A$ constitute the *format* of a particular pattern completion task. In the recall the memory should respond with *all* of the items $(x_1, \ldots, x_k)$ to *some* (sufficiently many) of the items of a stored pattern x.

Again the usual way of storing the information contained in a specific set S of the format $(P, k, P_1, \ldots, P_k, A)$ is simply to write it down in a storage medium organized into A lines each of which can store $\sum_i \log_2 P_i$ bits of information, namely $\log_2 P_i$ bits for each of the k items in a pattern $x = (x_1, \ldots, x_k)$.

In this procedure, search is again necessary, and again a certain amount of storage space is wasted for two reasons:

(a) Every number $\log_2 P_i$ has to be rounded up to the next larger integer.
(b) Not every combination of items may be possible, therefore P may be smaller than $\Pi_i P_i$, and thus $\log_2 P$ smaller than $\sum_i \log_2 P_i$.

Here, $A \log_2 P$ bits are stored but we need a (usually somewhat larger) storage space for $A \sum \log_2 P_i$ bits. For example, if we take as patterns English words of up to 10 letters, then $k = 10$, $\log_2 P \approx 14$ and $\log_2 P_i \approx 5$, thus $\sum_i \log_2 P_i \approx 50$.

Mathematically speaking, a *relation* is stored in the pattern completion task (a relation is a set of k-tuples, see also Apers 1983), and it is quite clear that the pattern mapping task can be regarded as a special case of it, since a mapping can be regarded as a special case of a relation.

3 The Structure of Associative Matrices or Networks

3.1 The Transmission of Activity

Let us start with the idea that some input activity is passed through a network of connections c_{ij} between the elements $e_1, \ldots, e_n$ and then read off from some or all of these elements. The simplest description for this flow of activity is the linear one:
If a_i denotes the activity of the element e_i at some moment in time, given as a real number, then after one time unit, the activity vector $a = (a_1, \ldots, a_n)$ has changed into $b = (b_1, \ldots, b_n)$, where $b_j = \sum_i a_i c_{ij}$ (i.e., $b = a \cdot C$).

The input activities x_j can simply be added onto the activities a_j of all or some of the elements e_j yielding

$$b = a \cdot C + x \quad \text{or} \quad b_j = \textstyle\sum_i a_i c_{ij} + x_j \ . \tag{1}$$

If not all elements e_j receive inputs (or $x_j \equiv 0$ for some e_j), we may call those e_j which do receive inputs the *input elements*. Similarly, we may call those elements e_j from which their output activity is read off, the *output elements*.

If an element does not receive any input from the network, it could be omitted without any effect unless it is used as an input element. Such elements e_j (satisfying $c_{ij} = 0$ for all i) will be called *pure input elements*. If an element does not give any output to the network, it could be omitted without any effect unless it is used as an output element. Such elements e_j (satisfying $c_{ji} = 0$ for all i) will be called *pure output elements*.

Many models of the flow of activity are somewhat more complex. In them b_j itself is not taken as the activity of e_j after one time unit, but a nonlinear function

$$\bar{b}_j = f(b_j) \tag{2}$$

is used instead. The function f is usually of a sigmoid shape; typical examples are $f(x) = \arctan x$ or the gaussian integral or the "threshold function" $f_\theta u(x) = 1$ for $x \geq \theta$ and $= 0$ for $x < \theta$.

Such a matrix or network can be used for the tasks of pattern mapping and pattern completion, if the input patterns (in both cases) are interpreted as (or coded into) special vectors of input activity $(x_j)_{j\,\mathrm{inputelement}}$ and the corresponding output activities $(b_j)_{j\,\mathrm{outputelement}}$ are interpreted as (or coded into) the output patterns (i.e. the completed patterns in the case of pattern completion). By the way, this representation suggests that in the case of pattern mapping the input elements should ideally be disjoint from the output elements, and both should be pure, whereas in the case of pattern completion all elements should ideally be taken as input elements and as output elements as well.

It should be noted that this framework is general enough to contain almost all mathematically formulated work on neural networks (at least as a discrete-time approximation), and associative memories. Even the usual (so called random access) computer memories (which map address patterns into content patterns) can be expressed in this framework: The addressing part can be said to code the address k into the vector $x_k = (\delta_{ik})$, which is given as an input to the storage part, which can be said to contain a matrix C, in the sense that each row of C is just the content of one address. The corresponding output will be $x_k C$, i.e. the kth row of the matrix.

3.2 The Learning Rule

If we want to use the connectivity matrix C for the tasks of pattern mapping and pattern completion as described above, it is clear that the matrix C has to contain the information necessary to perform this task for a specific list of (pairs of) patterns. Now the problem is how to set up the matrix to do the job. Here the basic idea is that the matrix should contain a superposition of the correlations of the (pairs of) patterns in the set S. More specifically the idea is to use something like

$$c_{ij} = \sum_{k=1}^{A} x_i^k \cdot y_j^k \quad \text{for a set} \quad S = \{(x^1, y^1), (x^2, y^2), \ldots, (x^A, y^A)\} \tag{3a}$$

and

$$c_{ij} = \sum_{k=1}^{A} x_i^k \cdot x_j^k \quad \text{for a set} \quad S = \{x^1, x^2, \ldots, x^A\}\ . \tag{3b}$$

One can imagine that this matrix is built up during a so-called adaptation-phase or learning-phase, where the connectivities c_{ij} are changed according to the above formulae.

In the case of pattern completion, assuming that every element is an input and an output element we could simply "teach" the matrix by giving the patterns x^k in the set S as inputs to it, leading to the activity $b = x^k$. This would require an adaptation rule of the type

$$\Delta c_{ij} \sim b_i \cdot b_j\ . \tag{3}$$

In the case of pattern mapping, when we assume that the input and output elements are disjoint and pure, we could "teach" the matrix a pair (x^k, y^k) in the set S by means of the same adaptation rule, if we could (at least during the learning phase) impose activity $b_j = y_j^k$ on the output elements, while giving the inputs x_i^k to the input elements.

As a matter of fact, almost all rules for the adaptation of the matrix c_{ij} that have been used in the literature on neural networks or associative memories, are of the type

$$\Delta c_{ij} = \alpha b_i \cdot b_j + \beta b_i + \gamma b_j + \delta \ , \tag{4}$$

where the values for the constants α,β,γ,δ may of course be different from model to model. It has been argued (e.g. Palm 1982a) that the product-nonlinearity in Eq. (4) is essential for a useful adaptation rule in this context. An additional saturation nonlinearity similar to Eq. (2) is also incorporated in many models: the new connectivity after one timestep is not simply given by

$$d_{ij} = c_{ij} + \Delta c_{ij}, \quad \text{but by} \tag{5}$$

$$\bar{d}_{ij} = g(d_{ij}) \tag{6}$$

with a sigmoid function g.

Now we have described the general framework in which we can discuss the modeling of neural networks, associative memories and even the more conventional computer memories (content addressable memories can also be discussed in this framework, e.g. Palm 1984).

The corresponding Eqs. (1–6) can be easily simulated on a conventional computer (but they usually take a long computation time) and it has been demonstrated in this way that the idea of using adaptive networks for the tasks of pattern mapping and pattern completion does indeed work reasonably well (e.g. Fukushima 1975, Kohonen 1977).

By this point we have already achieved something important. We have seen that scientists concerned with associative (computer) memories and those concerned with modelling of adaptive neural networks do work on the same mathematical subject. Moreover, the tasks formulated in the last section could be taken as at least one clearly formulated goal of neural modelling (as long as it is not directly concerned with the inputs to or the outputs from the brain). If we want to go beyond this point, if for example we want to discuss the relative advantages and disadvantages of different concrete versions of this general scheme, we need quantitative criteria to evaluate their performance.

4 The Information Storage Capacity as an Evaluation Criterion

As one important quantitative measure for the performance in the tasks of pattern mapping and pattern completion, I propose the *information storage capacity* (cf. Palm 1980). For these two tasks, the storage capacity can be defined operationally in the following way.

Pattern Mapping: Given the format (Pi, Po, A) we assume that the pairs (x, y) of patterns in the set S are collected at random from the finite set of all possible (x, y) – pairs. We calculate the information needed to specify one output y, which is $I(\text{output}) = \log_2 Po$. If the particular memory that we analyze, can produce errors with some probability, we try to estimate the average information $I(\text{error})$ that is still needed to correctly specify one output y, given the output of the memory. The information storage capacity of the memory is then

$$Ic = A(I(\text{output}) - I(\text{error})).$$

Pattern Completion: Given the format $(P, k, P_1, \ldots, P_k, A)$ we again assume that the patterns x in the set S are collected at random fro the finite set of all possible patterns. In many applications we can further assume that the "part" of a pattern that is presented to the memory in the recall, always contains a certain amount of information: $I(\text{part})$. Again we calculate the information contained in one pattern x as $I(\text{pattern}) = \log_2 P$, and take account of possible errors by estimating $I(\text{error})$. The resulting storage capacity of the memory is $Ic = A(I(\text{pattern}) - I(\text{part}) - I(\text{error}))$.

At first sight, this criterion might seem to be undebatable. There is, however, an alternative view, that appears quite naturally in the context of linear matrix relations like Eq. (1). When we use arbitrary vectors as our patterns on which the association ($x \to y$ or $x \to x$) has to be performed by means of the maxtrix C, then the number of possible patterns is infinite. Therefore the number of bits needed to describe a single association is also infinite and the storage capacity in terms of bits does not make sense, one should rather quantify it in terms of signal to noise ratio, or independence and number of the retrieved coordinates.

I did actually already discard this approach at the beginning of this paper, when I defined the format of the problems of pattern mapping and pattern completion in terms of, for example, the number of possible inputs. Let me briefly indicate a heuristic reason why.

Take the case of pattern mapping and a set $S = \{(x^1, y^1), \ldots, (x^A, y^A)\}$ of pairs of vectors x^i and y^i. Take C as an $m \times n$ matrix defined by Eq. (3a). If the vectors x^i $(i = 1, \ldots, A)$ are orthogonal $(\langle x^i | x^k \rangle = \delta_{ik})$, then we obtain

$$x^i C = \sum_k \langle x^i | x^k \rangle y^k = y^i \ . \tag{7}$$

This means a perfect retrieval of the outputs y^i, and it already turns out to be the optimal situation. Therefore, this approach typically leads to an optimal choice of $m = A$, whereas I am interested in situations where the size of the network $(m+n)$ can be small compared to the number of patterns stored (A).

Using the more "digital" approach of this paper and the information storage capacity as a criterion one can obtain optimal matrix configurations, where $m+n \ll A$ (for example for $m = n$ typical values for A are around $(n/\log n)^2$, see Palm 1980).

Let me explain this in some more detail. Consider again the problem of pattern mapping for a set $S = \{(x^1, y^1), \ldots, (x^A, y^A)\}$ of pairs of vectors x and y. Again we want to store this set in an $m \times n$ matrix C, but now we assume that the elements c_{ij} of C as well as the coordinates x_i and y_j of the vectors only have the values 0 or 1.

In this case c_{ij} is not defined by Eq. (3a), but by

$$c_{ij} = \left(\sum_{h=1}^{A} x_i^h \cdot y_j^h \right) \wedge 1 = \bigvee_{h=1}^{A} x_i^h \cdot y_j^h \tag{8}$$

($\wedge$ and $\vee$ signify the minimum- and maximim-operation, respectively).

Furthermore let us assume that the number of 1's in the vectors x^h and y^h is l and k, respectively, i.e. that

$$x^h \in \{0,1\}_l^m = \{(x_1, \ldots, x_m) : x_i \in \{0,1\} \quad \text{and} \quad \sum x_i = l\}$$

and $y^h \in \{0,1\}_k^h$.

If we want to retrieve the assignment $x^h \to y^h$ from the matrix C we can use the threshold function $f_\theta u$ (with $\theta = l$) in Eq. (2), i.e. we can form $z^h = f_l$ $(x^h \cdot C)$ and we know that z^h will contain the *ones* of y^h, but possibly some additional *ones*, which have to be regarded as errors. (If we are not sure that the input x^h is exactly one of the inputs in S, because it may itself contain one or two errors, we can try to work with $\theta = l-1$ or $\theta = l-2$ instead of $\theta = l$.)

For this procedure, I have optimized the information storage capacity in terms of the parameters k and l (for a sufficiently large $m \times n$ matrix). It turns out that the optimum is achieved, when about half the elements c_{ij} of the matrix are 1 (and the other half are zero).

Since one can estimate the number q of zeroes in the matrix C by

$$q = (1 - lk/mn)^A \approx \exp\left(-A \frac{lk}{mn}\right) \quad (\text{for } lk/(mn) \ll 1) \ , \tag{9}$$

one obtains

$$l \cdot k = (-\ln q)mn/A = (\ln 2)mn/A \ . \tag{10}$$

Also, one can estimate the average number n_e of errors (i.e. additional ones in the output z^h that are not contained in y^h) and the probability p_e of having at least one error, assuming $q = 1/2$:

$$n_e = (n-k)\cdot(1-q)^l = (n-k)\cdot 2^{-l} \text{ and } p_e = 1-(1-2^{-l})^{n-k} \ .$$

If n_e is not too large, let us say $n_e \approx \delta \leq 1$, i.e. $l \approx \log_2(n/\delta)$, one can estimate the information storage capacity as

$$\begin{aligned} Ic = A\cdot b = A\cdot\log_2\tbinom{n}{k} &\approx A\cdot k\cdot\log_2(n/k) = (\ln 2)(mn/l)\log_2(n/k) \\ &= (\ln 2)m\cdot n(\log_2 n - \log_2 k)/(\log_2 n - \log_2\delta) \approx (\ln 2)m\cdot n \ , \end{aligned} \tag{11}$$

for $1/\delta$, $k \ll n$.

Here $b = \log_2 Po = \log_2\binom{n}{k}$ denotes the information (in bit) contained in one output-vector $y^h \in \{0,1\}^h_k$.

Therefore, the size of the matrix is determined from the format (Pi, Po, A) by $m\cdot n\cdot\ln 2 = A\cdot\log_2 Po$, and l and k are determined by $l\cdot k = (\ln 2)m\cdot n/A = \log_2 Po$, and by the error criterion $l \approx \log_2(n/\delta)$.

Of course, these are only approximate equations, but the exact relations can also be calculated (see Palm 1980).

Some concrete examples for such configurations are given in Table 1. These and the following concrete matrix configurations are always optimized for *switching matrices* where the matrix coefficients can assume only two values, 0 and 1.

Another difference between the two approaches, let's call them the "*digital*" and the "*linear*" approach, is the following: for the linear approach the saturation nonlinearity introduced in Eq. (2) is regarded mainly as a nuisance, and therefore one usually tries to find arguments supporting the idea that the elements e_j stay in their linear "working range", i.e. b_j stays in the range where $f(b_j)$ is almost linear (also in the linear range of g in Eq. (6) one can erase a stored correlation simply by subtracting it, which is no longer possible near saturation). The digital approach makes active use of this nonlinearity as a cutoff to produce the desired digital output patterns. Moreover in my calcu-

Table 1. Some realistic matrix configurations (error probability < 0.001 for pattern mapping)

	m	n	A	Ic	$Ic/(m\cdot n)$
Pattern	1000	100	3,600	43,000	0.43
Mapping	1000	1000	25,000	466,000	0.466
	25,400	100	62,500	10^6	0.394
Pattern		1000	5,200	$3\cdot10^5$	0.3
Completion		1710	65,000	10^6	0.342
		17,000	$6.5\cdot10^6$	10^8	0.346

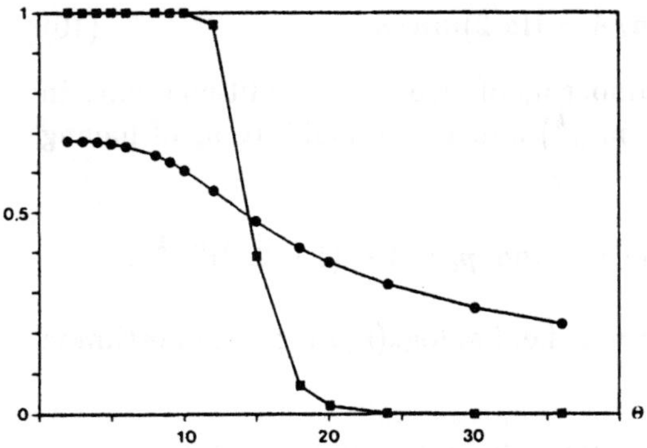

Fig. 1. For a fixed matrix configuration with $n = m = 10,000$, $A = 200,000$ and a constant density $d = 0.51$ of "ones" in the matrix, the error probability p_e(■) and the information storage capacity Ic (in 10^8 bits) (•) are plotted against the threshold θ

lations (Palm 1980) I have used the strongest threshold nonlinearity (f_θ) and have even regulated the threshold θ according to the input activity in order to minimize the error probabilities. Figure 1 shows the influence of the detection threshold θ on the error probability p_e and the storage capacity Ic.

In view of these calculations it turns out that even the simple digital version of an associative matrix memory, that I have analyzed in Palm (1980), is comparable in its storage capacity to conventional computer memories (see also Table 2, in which the memory requirement for the conventional storage configuration (each pattern in one line) is compared to one possible associative configuration for various problem formats).

Table 2

a) Pattern mapping

Format $Pi = 2^{64}$, $Po = 2^{16}$, $A = 62,500$, $Ic = 10^6$

Conventional memory matrix:	$62,500 \times 80 = 5 \cdot 10^6$
Associative memory matrix:	$25,400 \times 100 = 2.54 \cdot 10^6$

Format $Pi = 2^{50}$, $Po = 2^{200}$, $A = 5000$, $Ic = 10^6$

Associative memory matrix:	$3100 \times 1000 = 3.1 \cdot 10^6$

b) Pattern completion

Format $k = 10$, $P = 2^{14}$, $P_i = 2^5$ $(i = 1, \ldots, 10)$, $A = 2000$, $Ic = 12,000$

Conventional memory matrix:	$2000 \times 50 = 10^5$
Associative memory matrix:	$270 \times 270 = 72,900$

Format $k = 2$, $P = 2^{70}$, $P_2 = 2^{20}$, $A = 100,000$, $Ic = 5 \cdot 10^6$

Conventional memory matrix:	$100,000 \times 80 = 8 \cdot 10^6$
Associative memory matrix:	$3900 \times 3900 = 15.2 \cdot 10^6$

I think that a quantitative evaluation criterion, like the information storage capacity is essential not only for the comparison between associative memories and more conventional computer memories (as in Palm 1983, 1984) but also for the comparison of different models of associative memories or adaptive neural networks. In particular in the context of neural network modelling, the goals of the modelling were often not clearly defined and in many cases no criterion at all was given to judge the performance of the model, let alone a quantitative one. With the criterion introduced here, I think it should be possible to obtain quantitative answers to some questions concerning adaptive neural networks.

5 Associative Memory as a Model for Real Neuronal Networks

A model of an adaptive neural network usually contains two types of equations:

— equations for the flow of activity (like Eq. (1)), which have been termed "neuronic equations" by Caianiello (1961), and
— equations for the change of the connectivities, (like Eq. (3) or rather $\Delta c_{ij}(t) = a_i(t-1) \cdot b_j(t)$) termed "mnemonic equations" by Caianiello (1961); the latter might also be called synaptic equations.

A common assumption in neural network modelling (again made by Caianiello 1961) is that the mneomonic equations take place on a slower time-scale than the neuronic equations. Moreover, by reasonable stability assumptions (on the basis of inhibition and/or dynamical threshold control; see Palm 1981, Braitenberg 1978, Dammasch and Wagner 1984), upon a constant input, the neuronic equations should quickly fall into a steady state (or into a short cycle). Once the neuronal activity is in a steady state it will stay there until the input or some internal parameter like the threshold of the neurons is changed. During this steady state of neuronal activity, the mnemonic equations have time to work, to exactly the same effect as in the learning phase of an associative memory matrix for pattern completion.

Thus, the stabilized patterns of neural activity (that correspond to certain inputs) are "learned" in the sense of pattern completion and thereby further stabilized in such a way that later each of them can be activated by any sufficiently large part of itself. These stabilized self-completing patterns of neural activity are exactly what D.O. Hebb (1949) called *cell assemblies*. The mechanism by which they are "learned" is formulated in the mnemonic equations. In this way an adaptive neuronal network can be understood as an associative memory for the task of pattern completion.

There are basically two factual requirements on the connectivities in a real neuronal network needed for this interpretation as an associative memory.

(a) the connectivities between neurons must be variable according to a rule like Eq. (3),

(b) there must be sufficiently many such connections in the network such that almost any neuron can be connected to almost any other one.
In the following I want to elaborate a little on these two points.

(a) This is a requirement concerning the mnemonic equations. Mnemonic equations of the type of Eq. (4), where the change of synaptic efficiency depends only on the pre- and the postsynaptic activity have been called *local synaptic rules* (Palm 1982). Qualitatively they can be described by a simple table (see Palm 1982). It can be shown that such rules tend to form and stabilize assemblies, i.e., that they are *Hebb-like* (in the sense of Palm 1982) if and only if $\alpha > 0$ [in Eq. (4)].

In this context one question seems to be quite interesting: If we take a given neural network and look at local synaptic rules like Eq. (4) for a fixed $\alpha > 0$, what is the influence of the other parameters β, γ, and δ on the performance of the network in pattern completion, as measured for example by its information storage capacity?

Clearly an answer to this question requires many simulations and certainly also some additional theoretical work. We are presently engaged in this kind of work.

(b) In the human brain there are only of the order of 10^{14} synapses for some 10^{10} neurons. So it is clearly impossible to connect every neuron to every other one. In fact, the numbers seem to be far from this situation, since it would require some 10^{20} synapses.

This situation leads to a number of considerations. First of all, it is clear that the optimization results derived above do not directly apply to the brain. Still one may argue that there are ways of interpreting the real neuronal network of the cerebral cortex in terms of associative memory matrices. There could be many matrices realized in the brain (say, 10^4 matrices each containing 10^5 neurons and 3×10^9 synapses would account for 10^9 of the cortical neurons and 3×10^{13} of the synapses), or there could be other ways in which cortical connectivity comes close to the ideal of having every neuron connected to every other one.

For example it may be possible to connect (almost) every neuron to (almost) every other one via only a very few intermediates. Or it may be impossible to dissect the cortical network into two parts without cutting thousands of connections. To study these kinds of concepts and questions requires some mathematical work in the theory of graphs (which was initiated by Palm 1981) and some anatomical work on the global aspects of connectivity in real brains. In the mammalian cortex one can distinguish long-range and short-range connections, the latter forming a rather dense network of connectivities between neurons in the intermediate vicinity of each other (up to about 1 mm). The long-range connections, although distributed much more loosely, can obviously help a great deal to interconnect the cortical neurons more strongly, especially in the sense that each neuron can reach every other one in a very few steps, depending on the global organization of these connections. Now the concrete question is: do they really do so? This question has been studied anatomically (by H.R.P.-injections) for the mouse cortex by Horst Greilich in our laboratory. The actual arrangement of cortico-cortical long-range connections indeed makes it possible and conceivable that almost every neuron in the mouse cortex is connected to almost every other one via at most 3 intermediates (compare Fig. 2).

In any case, if one wants a rough idea of the possible size of a typical neural assembly or of the number of such assemblies in a large interconnected neural network like the cortex, one may take the following approach to apply the optimization results of this paper to real brains.

Given a network with n neurons and s synapses that is supposed to work like an associative memory for pattern completion, we may say that a particular synapse (connectivity) c belongs to a pattern (assembly) x, if the neurons i and j which it connects belong to x.

Following optimization arguments similar to those given in Sect. 3 and in Palm (1980), I will assume (somewhat arbitrarily but still quite reasonably) that for any particular synapse the probability of it belonging to at least one stored pattern, is about 1/2. This essentially distinguishes two kinds of synapses; quite effective ones that belong to some assembly and quite ineffective or maybe even non-existing ones (where two neurons just come close to each other without making a synapse) that belong to no assembly. With this assumption we obtain as in Eq. (10)

$$k^2 \approx \frac{n^2}{A} \ln 2, \quad \text{i.e. } k^2 A \approx n^2 \cdot \ln 2 \ , \tag{12}$$

where k is the size of a typical assembly and A the number of assemblies.

For the information storage capacity we may extrapolate from Eq. (11) the formula

$$Ic \approx s \cdot \ln 2 \ . \tag{13}$$

Obviously, this amount of information has to be shared by the A patterns or assemblies that are stored. If b is the number of bits stored in one assembly, we have

$$A \cdot b = Ic \approx s \cdot \ln 2 \text{ and } k^2 \approx b\frac{n^2}{d} \ . \tag{14}$$

If we wan to apply these formulae to a real brain we need an estimate of the information content of one typical assembly. One may get a rough idea of this number, if one identifies the information content of one activated assembly with that of a general short-term memory (while, of course, the whole information storage capacity Ic has to be interpreted as the information content of a long-term memory). For example, then, for $n = 10^{10}$, $s = 10^{14}$ and $b = 25$, one obtains $A \approx 4 \cdot 10^{12}$ and $k \approx 5000$.

These estimates for the number A of assemblies in the cerebral cortex and their size k should not be taken too seriously. The uncertainties in our

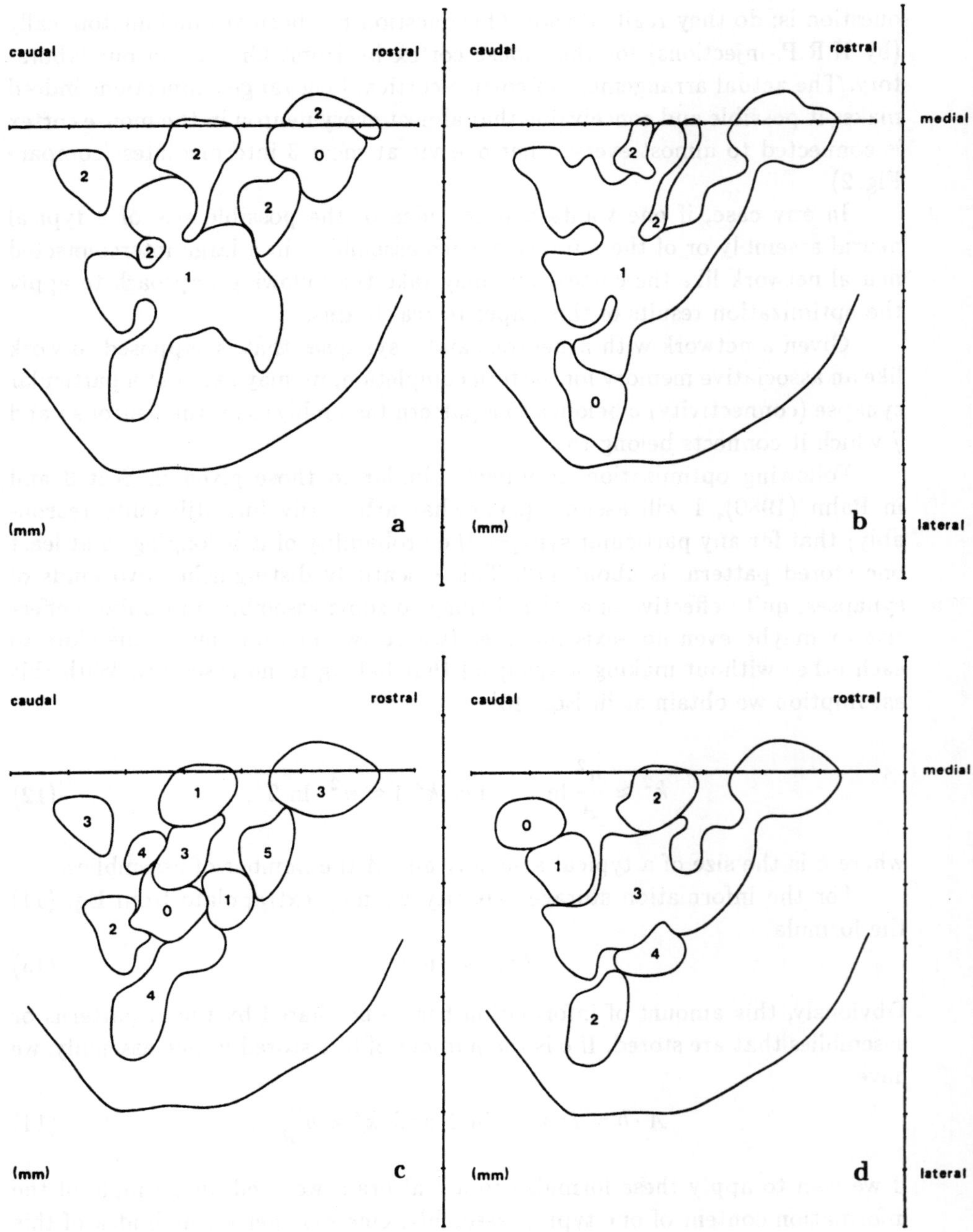

Fig. 2. Maps showing the number of steps needed to reach points in the upper half of the right cortical hemisphere through the cortico-cortical long range connections of the mouse. Zeroes indicate the starting points. For more details see Greilich (1984) or Greilich and Palm (1985)

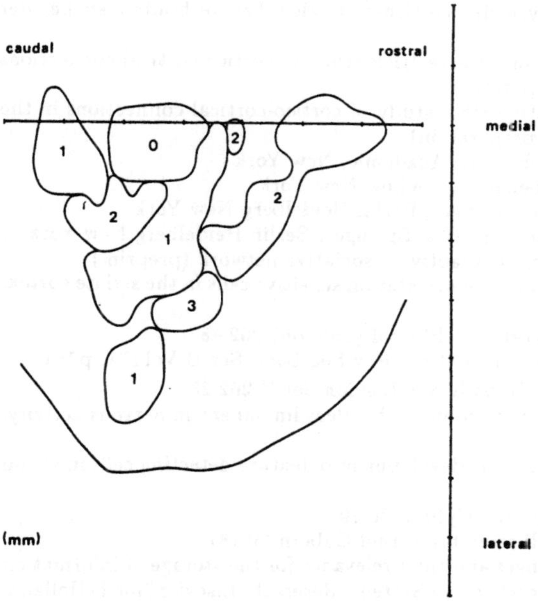

Fig. 2e

assumptions on n, s, b and $p = 1/2$ easily allow for changes of one order of magnitude up or down for A and about half that much for k. Furthermore, the idea of threshold regulation of the succession of assemblies in the cortex (as explained in Chap. 12 of Palm 1982) implies that there should be assemblies of different sizes at different threshold levels, i.e. there should be a whole "working range" of normal assembly sizes. Since we have derived the value for b from estimates on the maximal information content of a general short-term memory, k should be regarded not so much as the average size of an assembly, but rather as the upper margin of this "working range" (which might then be something like 1000–5000 neurons per assembly).

References

Amari S-I (1974) A method of statistical neurodynamics. Kybernetik 14:201

Anderson IA (1972) A simple neural network generating an interactive memory. Math Biosci 14:197

Anninos PA, Beek B, Csermely TJ, Harth EM, Pertile G (1970) Dynamics of neural structures. J Theor Biol 26:121

Apers (1983) Query processing and data allocation in distributed database systems. Math Centrum, Amsterdam

Braitenberg V (1978) Cell Assemblies in the Cerebral Cortex. In: Heim R, Palm G (eds) Theoretical approaches to complex systems. Springer, Berlin Heidelberg New York, p 171

Caianiello ER (1961) Outline of a theory of though processes and thinking machines. J Theor Biol 1:204

Dammasch IE, Wagner GP (1984) On the properties of randomly connected McCulloch-Pitts Networks: Differences between input-constant and input-variant Networks. Cybern Syst 15:91

Fukushima K (1975) Cognitron: a self-organizing multilayered neural network. Biol Cybern 20:121

Greilich H (1984) Quantitative Analyse der cortico-corticalen Fernverbindungen bei der Maus. Dissertation, Tübingen

Greilich H, Dortenmann M (1985) A quantitative HRP-study of cortico-cortical connections in the mouse. Anat Aspects (in preparation)

Greilich H, Palm G (1985) A quantitative HRP-study of cortico-cortical connections in the mouse. Functional implications (in preparation)

Griffith JS (1971) Mathematical Neurobiology. Academic, New York

Hebb DO (1949) The organization of behaviour. Wiley, New York

Kohonen T (1977) Associative memory. Springer, Berlin Heidelberg New York

Kohonen T (1980) Content addressable memories. Springer, Berlin Heidelberg New York

Lansner A (1983) Pattern processing in an adaptive associative network (preprint)

Malsburg C von der (1973) Self-organization of orientation sensitive cells in the striate cortex. Kybernetik 14:85

Marr D (1969) A theory of cerebellar cortex. J Physiol (London) 202:437

Marr D (1970) A theory of cerebral neocortex. Proc Roy Soc Lond Ser B Vol 171, p 161

Marr D (1971) Simple memory. Philos Trans R Soc London Ser B 262:23

McCulloch WS, Pitts W (1943) A logical calculus of the ideas immanent in nervous activity. Bull Math Biophys 5:115

Nass MM, Cooper LN (1975) A theory for the development of feature detecting cells in visual cortex. Biol Cybern 19:1

Palm G (1980) On associative memory. Biol Cybern 36:19

Palm G (1981) Towards a theory of cell assemblies. Biol Cybern 39:181

Palm G (1982a) Rules for synaptic changes and their relevance for the storage of information in the brain. In: Trappl R (ed) Cybernetics and Systems Reserach. Elsevier North-Holland, Amsterdam New York

Palm G (1982b) Neural assemblies. An alternative approach to artificial intelligence. Springer, Berlin Heidelberg New York

Palm G (1984) Local synaptic modification can lead to organized connectivity patterns in associative memory. In: Frehland E (ed) Synergetics: From microscopic to macroscopic order. Springer, Berlin Heidelberg New York

Steinbuch K (1961) Die Lernmatrix. Kybernetik 1:36

Uttley AM (1956) Conditional probability machines and conditioned reflexes. In: Shannon CE, McCarthy I (eds) Automata studies. Princeton Univ Press, Princeton

Wilson KV (1980) From associations to structure. Elsevier North-Holland, Amsterdam New York

Warren McCulloch and Walter Pitts: A Logical Calculus of the Ideas Immanent in Nervous Activity

G. PALM[1]

The classical paper by McCulloch and Pitts on "a logical calculus of the ideas immanent in nervous activity" had an enormous impact on the development of brain theory in the broadest sense. It appeared in 1943 and was the starting point for many theoretical investigations up to the present day: its basic idea was that the activation of a neuron inside a brain stands for the actual truth of a proposition about the outside world. Elementary propositions about the outside world are verified through sensors. The neurons to which these sensors are connected may themselves represent more complicated combinations of these propositions. Since it is possible to implement the logical connections *not*, *and*, and *or* by means of neural connections and appropriate thresholds of the neurons, one can represent *every* conceivable *finite* logical combination of the elementary propositions in a neural network.

This idea of representation of relevant propositions about the outside world in single neurons has been the basis for the leading experimental paradigm in neurophysiology from the 1950's up to the present day, starting with the work of Lettvin et al. (1959) it has led to the breakthrough of Hubel and Wiesel (e.g. 1963, 1977) and to the concepts of the receptive field – and the grandmother neuron. In a sense, McCulloch was the grandfather of the grandmother neuron.

But this representation of every finite logical combination of elementary sensory propositions is already possible in a hierarchical neural network without feedback, where the neurons are at most two steps away from the input. So the main part of the paper is concerned with those infinite combinations of elementary propositions that can be represented in the single neurons of a network with feedback – and those that cannot. Here the results of the paper become cryptic, but the question has turned out to be a fruitful one, and many important studies in the field of automata theory, notably Kleene (1956), have been devoted to the exact characterization of the class of all propositions that can be represented in certain types of neural networks or automata.

The exact definition of a neural network, that was necessary for the logical formulation in the paper, has led to many studies on networks of formal neurons, where the formalization of neural properties ranges from the most formal

[1] Max Planck Institut für Biologische Kybernetik, Spemannstraße 38, 7400 Tübigen, FRG

Brain Theory
Edited by G. Palm and A. Aertsen

threshold nerons or McCulloch and Pitts neurons to a detailed mathematical description close to the famous Hodgekin-Huxley equations (see MacGregor and Lewis 1977).

But for me the paper by McCulloch and Pitts is also interesting from a psychological or sociological point of view. As a mathematician who is used to communication with biologists and neurologists I have become used to making hard compromises between mathematical rigour and readability in the formulation of arguments, since I usually try to make the essence of my ideas intelligible not only to mathematicians but also to biologists. McCulloch and Pitts are breathtakingly arrogant in their use of mathematical formalism. Without any consideration for the mathematical background of their prospective readers they use the uncommon logical formalism of Carnap (1938) and Hilbert and Ackermann (1927) for the presentation of their results. This makes the mathematical portions of the paper, i.e. almost all of it, quite unreadable even for mathematicians. For example, Kleene (1956) remarks: "we found the part of their paper dealing with arbitrary nerve nets (Part III of McCulloch and Pitts, 1943) obscure, so we have proceeded independently there." (p. 4) and in a footnote on p. 17, after giving a counterexample to one of the statements in the paper he says: "This apparent counterexample discouraged us from further attempts to decipher Part III of McCulloch and Pitts, 1943."

The use of logical formalism by McCulloch and Pitts is a clear consequence of their conviction that in physiology the fundamental relations are those of two-valued binary logics. But this conviction is not shared by most biologists and the formalism itself is so discouraging that the whole paper cannot have been read and understood by more than a very few people. Still, the paper has had an enormous influence and has been cited in various theoretical studies.

I think this story should encourage us theoreticians to use the mathematics that we know more freely, if we have something important to say. The essence will be understood, if it is there.

References

Carnap R (1938) The logical syntax of language. Harcourt-Brace, New York

Hilbert D, Ackermann W (1927) Grundzüge der theoretischen Logik. Springer, Berlin

Hubel DH, Wiesel TN (1963) Receptive fields of cells in striate cortex of very young, visually inexperienced kittens. J Neurophysiol 27:994

Hubel DH, Wiesel TN (1977) Functional architecture of macaque monkey visual cortex. (Ferrier Lecture) Proc R Soc London Ser B 198:1

Kleene SC (1956) Representation of events in nerve nets and finite automata. In: Shannon CE, McCarthy J (eds) Automata studies, vol 34. Princeton Univ Press, Princeton

Lettvin JY, Maturana HR, McCulloch WS, Pitts W (1959) What the frog's eye tells the frog's brain. Proc IRE 47:1940

MacGregor RJ, Lewis ER (1977) Neural modelling. Plenum Press, New York

McCulloch WS, Pitts W (1943) A logical calculus of the ideas immanent in nervous activity. Bull Math Biophys 5:115

Donald Hebb: The Organization of Behavior

G.L. SHAW[1]

I consider this a great privilege to be able to briefly remark on D.O. Hebb's marvellous book "Organization of Behavior: A Neuropsychological Theory" which he wrote in 1949. Hebb's ideas have had a profound influence on brain theory, in particular his famous "A Neurophysiological Postulate" governing the correlated pre-post synaptic changes which are the basis for the engram or memory trace. Although, there are *many* different forms of Hebb's postulate, I believe that essentially all "viable" mammalian cortical models embody some version of his idea: "Let us assume then that the persistence or repetition of a reverberatory activity (or "trace") tends to induce lasting cellular changes that add to its stability. The assumption can be precisely stated as follows:

When an axon of cell A is near enough to excite cell B and repeatedly or persistently takes part in firing it, some growth process or metabolic change takes place in one or both cells such that A's efficiency, as one of the cells firing B, is increased".

Since the axon of each cortical neuron branches to form roughly $10^3 - 10^4$ synapses with other cortical neurons, a pre-post correlation algorithm for synaptic change would be of enormous use to the system. (This might be contrasted to the *much* smaller branching each neuronal axon has in invertebrate ganglia.)

Although *firm* evidence for the "Hebb-synapse" has not been found experimentally, I believe that further examination of Hebb's insightful remarks might be of considerable help in designing future experiments, in particular his statements on cell assemblies. The role of cell assemblies (of a smaller size than Hebb's) in cortical function has been one of the principal themes in my research. Whenever, I get a "new" idea I go back and reread parts of "The Organization of Behavior" and I am usually surprised by how much I learn. Thus let me select out a few pertinent paragraphs on assemblies among those I had underlined in my well-read copy.

[1] Physics Department, University of California, Irvine, CA 92717, USA

Brain Theory
Edited by G. Palm and A. Aertsen

At the beginning of Chap. 4 he states "It is proposed first that a repeated stimulation of specific receptors will lead slowly to the formation of an "assembly" of association-area cells which can act briefly as a closed system after stimulation has ceased; this prolongs the time during which the structural changes of learning can occur and constitutes the simplest instance of a representative process (image or idea)." Then shortly, he refers to "a growth process accompanying synaptic activity" which differs from earlier suggestions in several aspects including "structural connections are postulated between single cells, but single cells are not effective units of transmission". Then concentrating on the crucial question of maintaining reverberatory activity for some fraction of a second or longer (the "duration of a single 'conscious content' ") he states "Lorente de Nó (1938) has put stress on the fact that activity in a short closed circuit must be rapidly extinguished and could hardly persist as long as a hundredth of a second." He then presents a new idea in Fig. 10 in which arrows represent not activity between neurons but "multiple pathways of whatever complexity is necessary so that each arrow stands for a functional unit.... The sort of irregular three-dimensional net which might be the anatomical basis of perceptual integration.... This then is the cell-assembly."

He then emphasizes the unlikely exactness of connections: "The answer of course is statistical". *Also,* for *timing*: "Once more, the oversimplification... is highly misleading. At each synapse there must be a considerable dispersion in the time arrival of impulses, and in each individual fibre a constant variation of responsiveness; and one could never predicate a determinate pattern of action in any small segment of the system. In the larger system, however, a statistical constancy might be quite predictable." Hebb's ideas here, I believe, are extremely relevant today. In the cortical organization principle put forward recently by V. Mountcastle, the cortical column is comprised of irreducible processing subunits, and sophisticated spatial-temporal patterns of activity are present. The key questions are what are the appropriate spatial and temporal scales in which to look for such patterns or "coding" of information.

Hebb, of course, does not stop with his discussion of the "Growth of the Assembly" in Chap. 4 with which most neural modeling has been concerned, but launches into new powerful ideas starting in Chap. 5 entitled "Perception of a Complex: The Phase Sequence". To discuss perception, he stresses that "what we are aiming at here is the solution of a psychological problem. To get psychological theory out of a difficult impasse, one must find a way of reconciling three things without recourse to animism: perceptual generalization, the stability of memory, and the instabilities of attention. As neurophysiology, this and the preceding chapter go beyond the bounds of useful speculation. They make too many steps without experimental check. As psychology, they are part of a preparation for experiment.... Although this discussion is not anchored with sufficient neurophysiological evidence, it still has a compass. It is guided throughout by the necessity of conforming to the concrete evidence of behavior." His discussion of the temporal sequences of activity involved in perception seem to me to be of profound importance. I plan to return to these ideas of Hebb at some stage in my work when I am considering these problems. Indeed, although enormous progress in the past 35 years has been made in the understanding of brain function and in the gathering of anatomical, physiological and psychological data, Hebb's book remains of great relevance.

Reference

Hebb DO (1949) The organization of behavior. Wiley, New York

Alan Turing: The Chemical Basis of Morphogenesis

W.J. FREEMAN[1]

Alan Turing should have been a national hero in Great Britain after the Second World War; by his unique accomplishment in solving the code for a captured German encrypting device called "Enigma", he saved more lives and treasure than any other of his countrymen. Yet his work was "Top Secret" and slated to remain so for another generation; he received no reward or recognition. To the contrary his then unacceptable life style and his naivete in social and political affairs led him afoul of British law. Faced with the choice between prison and pharmaceutical castration he accepted the latter and within a year was dead from the taste of an apple he had laced with cyanide, a macabre re-play of the fairy tale of Snow White.

His accomplishments in founding the field of artificial intelligence are legendary. Examples are the Turing machine, the universal automaton that can solve any logical problem or perform any logical task short of the self-referential, and the Turing test, which is widely quoted as the touchstone for recognizing artificial intelligence once it has been created. He did this work during the 1940's, but became bored with the field and irked by the constraints imposed by security. More than this, he became fascinated with the phenomenon of the emergence of order in biological systems, beginning with the fertilized ovum and proceeding to the formation of symmetrical structures in animals and plants. In the space of the last two years of his fore-shortened life he wrote and published this remarkable paper on the mathematical analysis of the manner in which spatial and temporal structure (morphogenesis) might be explained as a property of chemical systems.

At the time this was rank heresy. Chemistry did not explain or attempt to explain such phenomena, although Benard cells and Liesegang rings had been well known for half a century. The work lay unremarked in the literature for a decade, until the Zhabotinsky-Belousov reaction was discovered and widely publicized; with this stimulus from the experimental realm theorists such as Katchalsky, Eigen, Prigogine, Thom, and others undertook further explorations of the emergence of order from uniformity and from chaos, and of the kinds of nonlinear, distributed systems that give rise to these self-organized patterns

[1] Department of Physiology-Anatomy, University of California, Berkeley, CA 94720, USA

Brain Theory
Edited by G. Palm and A. Aertsen

(also known as "dissipative structures"). At present the study of nonlinear dynamics has become a major new area of mathematics, with applications in such diverse fields as plasma physics, hydrodynamics, meteorology, economics, sociobiology and developmental biology. The introduction of this new way of looking at biological systems is just beginning for studies of brain function and behavior; it is likely that it will become one of the most important conceptual tools for neurobiologists in the coming decade, if not the single most valuable pathway to follow in unravelling the complexities of large scale neural dynamics.

Moreover, the explanation of brain function most likely to succeed will incorporate an understanding of dynamic brain function in the adult with comprehension of the embryological growth and development of the fetal brain, because the processes by means of which function controls the emergence of form in the fetus are likely to be basically the same as those that subserve learning in the adult. The foundation that Alan Turing laid may then become as important in neuroscience as the Turing machine has been for the field of AI; and it may lead to forms of AI that present followers of Turing have not yet dreamt of.

Reference

Turing AM (1952) The chemical basis of morphogenesis. Philos Trans R Soc London Ser B 237:37–72

W.G. Walter: The Living Brain

W.J. FREEMAN[1]

This book is not a research monograph focused on brain theory in a heavy manner; it is a delightful collection of essays that rambles through Grey Walter's experiences, ideas and inventions in the classical age of vacuum tube ("valve") electronics. He was a bright young engineer in the 1930's when the electroencephalogram became accessible and widely known; his major contributions included correct localization of the alpha rhythm to the occipital cortex, instead of to the frontal cortex as proposed by its discoverer, Hans Berger in 1929, and the detection of slow waves and their significance for the diagnosis of brain tumors and abscesses. After the Second World War he gathered together some of the electronic "wizards" extolled by Churchill in his History of the War, together with surplus radar and other electronics equipment, and turned his laboratory in Bristol, England into one of the world's leading centers for EEG research. Here he developed his "toposcope", a device for displaying the phase patterns of the alpha rhytm recorded on the scalp.

His most remarkable accomplishment lay in the development of a set of mechanical tortoises that became known as "Grey Walter's turtles". These were three-wheeled toy vehicles that moved themselves in cycloid fashion with two electric motors. Each had a shell that was sensitive to touch, so that if it bumped into another object, it would stop, back up, and go around it. Each had a photocell on its carapace and a small signal light that turned on whenever its motors were active. A simple set of two to four mechanical relays comprised its "brain". The device was set to seek a medium level of illumination with its photocell. Its hutch, wherein it could plug itself in to recharge its battery, was brightly lit, so when it was released, it moved out into the room and away from the light. As its battery ran down, its sensitivity decreased, so that it returned to its hutch and was recharged. If it encountered a mirror, it stopped to reorient to the light, but the light went out and its started to circle again, only again to stop briefly. If it encountered another of its own kind, a stately dance ensued, with bumping and backing.

[1] Department of Physiology-Anatomy, University of California, Berkeley, CA 94720, USA

Brain Theory
Edited by G. Palm and A. Aertsen

These were the simplest of the series, and they displayed goal-seeking behavior relating to both approach and avoidance, as well as behavior characteristic of self-recognition and recognition of others, all with two sensors, two effectors and a small number of relays. Walter built progressively more advanced models that were capable of learned behavior; for example, his device could learn to go around an obstacle in its path toward the hutch, and if the obstacle were removed on a later trial, the device went around the site of the obstacle no longer there.

The point of these devices was to illustrate the complexity of behavior that could be elicited from the simplest automata with sensors and effectors incorporated into feedback loops. His idea was that brains appear to be immensely complex because of the large numbers of neurons and their interconnections, but that if they were viewed in the right way in terms of a much smaller number of assemblies of neurons, then the analysis of their function in relation to behavior would be feasible in the context of cybernetics and systems theory. His devices were the first and are still the only free-ranging robots capable of exploring a limited environment and learning their way around. It is a curious fact that his turtles were more laughed at than admired or understood in his time, and that they have been almost totally forgotten. Their message is still clear and apt: the brain and its behavior are not as complex as they seem; we have been seeing them from the wrong perspective. These turtles are in the nature of an existence theorem, that tells us that answers lie at hand to our questions of how the brain works, and tells us something about how we should formulate questions we are asking.

Reference

Walter WG (1953) The living brain. Norton, New York

John von Neumann: The Computer and the Brain

W.J. FREEMAN[1]

"Johnnie" von Neumann lived and worked among the brightest and most accomplished physicists and mathematicians of his era, the age of quantum mechanics and the "The Bomb", to both of which he was a major contributor; he was no one's peer, and was widely known as the man who was never bested in speed and accuracy of thought except by ruse. When the war was over and Hiroshima lay in ruins, he turned his talents to work on the newly emergent digital computer, recognizing instantly the latent potential of this still ungainly machine. He adopted whole heartedly from W. McCulloch and W. Pitts the language and concepts of the nervous system as a digital biological machine, and he used these metaphors liberally to guide his thinking in adapting the digital devices to exploit Booleam algebra and the "Theory of Games". He was a master politician of science and a uniquely effective salesman for the utility of "Gamesmanship" in human affairs ranging from "Real-politik" to corporate business and the operation of railroads, all done best by the computer. His greatest accomplishment was in devising a method for encoding in symbols the sets of instructions for operations on data; thereby with the invention of programming he created the first truly universal automaton. Programmers still express ambivalence about the immense utility of his artifice when they complain about the strait-jacket imposed by the "von Neumann heuristic". He was the Godfather of FORTRAN and all other commonly used programming languages.

In the mid 1950's when at the height of his scientific and political power he accepted an invitation to give the Silliman Lectures at Yale University on the subject that was to become the title of his book. Shortly thereafter he developed an osteosarcoma; as his health failed he withdrew progressively until all his remaining energy went to the completion of the lectures; his widow gathered the incomplete fragments and published them posthumously.

His message is among the most poignant to be found in scientific history. Over and over he compares the properties of computers with the properties of neurons and the brain and describes unequivocal differences: digital vs. analog, serial vs. parallel; arithmetic depth vs. logical depth; and so on. He concludes

[1] Department of Physiology-Anatomy, University of California, Berkeley, CA 94720, USA

Brain Theory
Edited by G. Palm and A. Aertsen

that the brain and the computer are fundamentally dissimilar, and that whatever the primary language of the brain might be, "it cannot fail to differ considerably from what we consciously and explicitly consider as mathematics" (p. 82). His metaphor had failed, and he seemed to stand in awe before the biological machine that he had tried to recreate, but that in the end he realized he did not comprehend.

Reference

J Neumann von (1958) The computer and the brain. Yale Univ Press, New Haven

Eduardo Caianiello: Thought Processes and Thinking Machines

P. JOHANNESMA[1]

In 1960 E.R. Caianiello of the "Istituto di Fisica Teorica dell'Universita di Napoli" submitted a paper to the *Journal of Theoretical Biology* with the title "Outline of a Theory of Thought Processes and Thinking Machines." While the author tried to remain modest in naming his paper only "Outline of a Theory", he demonstrated his optimism in relating "thought" and "processes" as well as "thinking" and "machines". This optimistic attitude reflected probably both the age of the author, around 35 at that time, as well as the general scientific atmosphere of that period.

Main precursors of the ideas of Caianiello were McCulloch and Pitts in 1943 with "A Logical Calculus of the Ideas Immanent in Nervous Activity" and Beurle in 1956 with "Properties of a Mass of Cells Capable of Regenerating Pulses". The widely known but sparsely read paper of McCulloch and Pitts stressed the logical and computational aspects of neural nets, especially their universal computing capacity, while Beurle focused upon statistical forms of information processing and associative learning.

As a summary of the ideas presented by Caianiello and an illustration of his attitude with respect to mind-brain we reproduce here part of the abstract to his paper.

"Thought-processes and certain typical mental phenomena are schematized into exact mathematical definitions, in terms of a theory which, with the assumption that learning is a relatively slow process, reduces to two sets of equations: "neuronic equations", with fixed coefficients, which determine the instantaneous behavior, "mnemonic equations", which determine the long-term behavior of a "model of the brain" or "thinking machine". A qualitative but rigorous discussion shows that this machine exhibits, as a necessary consequence of the theory, many properties that are typical of the living brain: including need to "sleep", ability spontaneously to form new ideas (patterns) which associate old ones, self-organisation twoards more reliable operation, and many others."

[1] Department of Medical Physics and Biophysics, University of Nijmegen, Nijmegen, The Netherlands

Brain Theory
Edited by G. Palm and A. Aertsen

The epistomological base of Caianiello's mathematical description is stated clearly in the introduction.

"Our main guiding principle has been the conviction, strenghtened by these discussions, that the human brain, tremendous in its complexity, yet obeys, if one looks at the operation of individual neurons, dynamical laws that are not necessarily complicated; and that these laws are such as to engender in large neuronal assemblies collective modes of behavior, to which thought-processes are correlated."

On the other hand the goals are defined as:

"A convenient formulation of these laws appears therefore as the primary objective of a research of this nature; it can only be achieved by trial and error by the process, familiar in the physical sciences, of abstracting what seems relevant into a simplified model of the real thing."

and moreover as:

"Our equations are also intended to provide a useful tool both for theoretical study and for experimentation in this respect."

The formulation of the equations for neural information processing are preceded by an "adiabatic learning hypothesis" assuming the separation of variables and parameters by a difference in time-scale. The "Neuronic Equations" are given by:

$$u_h(t+\tau) = 1\left[\sum_k \sum_r a_{hr}(t, r\tau) u_k(t - r\tau) - s_h\right]$$

where

τ is discretisation of time,
s is threshold
u is binary action-potential
a is synaptic coupling.

These equations represent the neural interaction: a step function compares the linear spatio-temporal summation of action potentials with a threshold and decides if either a zero or one is created at output variable. This approach is completely in the McCulloch and Pitts' tradition of formal dynamical logic.

The complementary set of equations is based on Hebb's idea of association by synchrony. The "mnemonic equations" are (in a slightly modified form):

$$\frac{d}{dt} a_{hk}(t, r\tau) = \alpha(r) u_h(t) u_k(t - r\tau) - \beta(r) a_{hk}(t, r\tau).$$

These equations represent the change of the synaptic connections: structural evolution based on the actual process.

The formulation of these two sets of equations supplies a firm base of signal processing in associative computing nets; it also formed an acceptable approximation of the information processing in the brain. The question remains however, how relevant the equations are for description and understanding of thinking and learning. In the qualitative discussion, following the formulation of the equations, Caianiello writes:

"We define a thought process, operationally, as a solution of the neuronic Equations, or, equivalently, as the corresponding "motion" in the machine."
and somewhat further on he states:
"The systematic algebraization of logic, which is the real content of the N.E...., permits us to pose for them much more general questions, ...; an appropriate treatment of the Neuronic Equations will permit, as will the equations of motions of dynamical systems, the search for those long-range collective solutions which, in our scheme, form the basis for a useful analysis of thought-processes".
And again in the context of the mnemonic equations:
"...; it was also implicitly assumed that the machine is indifferent to the "meaning" (referred to any standards) of what it knows genetically or has learnt during its past activity."

In the discussion of the equations Caianiello describes processes which may take place in the machine (stability, reverberations, parameter changes) as well as possible functions of the machine (pattern analysis, generalisations, abstraction).

At the end of the paper the author gives, anno 1961, his "Further Outlooks":

"As a final comment, we think it appropriate to remark that the general formalism of N.E., M.E., and A.L.H. which is expounded here, seems to us to admit of a far wider range of applicability than that to which it has been restricted in this work. The N.E. in fact, for instance, serve only to express in a more or less schematic manner, the fact that a decision is taken, after a weighted evaluation of the information which lasts a finite time, by a member of a set, and that such a decision is bound to affect other decisions, etc. We may change their name into that of decision equations, call the M.E. evolution equations, and take all our considerations over to the study of social or economical or other collective phenomena.

We have pursued this line of thought in several directions for personal amusement, and have soon found, to our surprise, that the qualitative analysis given here for thought-processes applies as well, mutatis mutandis (that is, names), to a great many other instances. We believe that, as soon as it becomes possible to agree on a concrete choice of schemes and numbers, quite reasonable predictions may be made in this way about, say, the operation of a stock-exchange, the variation in time of a parameter in feminine fashion, the type of national government that would best obey Wiener's principle, and so on. This we say with at least the same degree of assurance that we have found in the economists who apply the Schrödinger equation to the study of their problems.

Although we have refrained here from a quantitative analysis of the several mathematical problems formulated in the course of this work, the results we have already obtained in this direction seem to justify some optimism; if these expectations are not illusory, then the present formalism might help us to gain a finer knowledge of some physical phenomena that can now be treated only with statistical methods."

How is the situation anno 1985? In a long series of papers from 1960 onward Caianiello and co-workers developed the theory and derived many properties of the processes in the neuronic machine. To this meeting he promised the complete solution of the neuronic equations: the fulfillment of this promise is demonstrated in his contribution to these proceedings. As a consequence we know much about the mathematics of interconnected threshold elements, parallel computing and logic nets. For brain theory it may be useful because it does supply a well-documented mathematical metaphor of the brain. Shortcomings of the equation for this purpose are the discretisation of time and the absence of stochastic aspects. On the other hand they may be very useful for an investigation of the relation between correlation of neural activity as measured from the process and connectivity of neurons as present in the structure. It may well be that some problems present in the analysis of muli-unit recordings can be solved with the help of the theoretical work of Caianiello et al.

A fundamental point for the evaluation of any theory of thought processes is the relation of process and function. In the 1961 paper Caianiello equates these two in his operational definition of thought-processes and eliminates any "meaning" out of his machine. From the present point of view we might say that he did not discriminate between fact and process which belong to the syntactical level and function and meaning which are on the semantical level. In several of the contributions to this meeting, including the one of Caianiello, the struggle can be witnessed to find a scientific description of the mental aspects of the brain. The concept of a "representational system" forms part of this endeavour.

References

Beurle RL (1956) Properties of a mass of cells capable of regenerating pulses. Phil Trans Roy Soc London A204:55–94

Caianiello ER (1961) Outline of a theory of thought processes and thinking machines. J Theor Biol 2:204–235

Hebb DO (1949) The organisation of behavior. A neuropsychological theory. Wiley, New York; Chapman and Hall, London

McCulloch WS, Pitts WH (1943) A logical calculus ofideas immanent in nervous activity. J Math Biophys 5:115–133

Frank Rosenblatt: Principles of Neurodynamics: Perceptrons and the Theory of Brain Mechanisms

C. VON DER MALSBURG[1]

Frank Rosenblatt's intention with his book, according to his own introduction, is not just to describe a machine, the perceptron, but rather to put forward a theory. He formulates a series of machines. Each machine serves to introduce a new concept.

The basic model is a three-layer perceptron. Each layer is a set of simple threshold elements, or "neurons". Layers S, A, R are coupled in series by synaptic connections $S \rightarrow A \rightarrow R$. S is a sensory surface (e.g. a retina). Layer A contains "associator units", or feature detector cells, and R contains R-cells, or "recognition cells". Present a pattern S_1 in S. After one synaptic delay a pattern A_1 in A is active, after another delay a set of R-cells fires. An R-cell is to be activated precisely when the pattern projected onto S is of a certain type (e.g. is a triangle). Each A-unit receives connections from a specific subset of the cells of S and is fired by a specific pattern on this support. The purpose of the A-layer is to reduce the overlap between patterns which are to be classified differently.

The weights of the connections from A to an R-cell are set by synaptic plasticity. When a pattern has been presented to S, an external "teacher" decides for the R-cell whether it has responded correctly. If the cell fires although it shouldn't, the synaptic weights of all currently active A-cells to the R-cell are reduced. If the cell doesn't fire although it should, the synaptic weights of all active A-cells to the R-cell are increased. No changes take place in the case of correct response.

There is a convergence theorem. For the case that there exists a vector of synaptic $A \rightarrow R$ connections which leads to the correct classification of patterns by the R-cell, the theorem states that such a vector can be found by the above learning procedure in finite time.

It is a serious weakness of the three-layer perceptron that it cannot generalize. If an R-cell has learnt to discriminate a particular pattern in S from all other patterns, it does not recognize the same pattern in other positions on S. Rosenblatt therefore introduces the four-layer perceptron. It has layers S, $A^{(1)}$, $A^{(2)}$ and R. For each cell in $A^{(2)}$ there exist many cells in $A^{(1)}$, each

[1] Max-Planck-Institut für Biophysikalische Chemie, 3400 Göttingen, FRG

Brain Theory
Edited by G. Palm and A. Aertsen

responding to the same feature in a different position on S, each having a connection to the cell in $A^{(2)}$, and each being able to fire the cell in $A^{(2)}$. Thus, a cell in $A^{(2)}$ responds to a feature irrespective of the feature's position. The set of cells in $A^{(2)}$ activated by a pattern in S is invariant with respect to the position of the pattern in S. (The concept of invariance is treated by Rosenblatt with respect to more general groups of transformations.)

With later versions of the perceptron Rosenblatt introduces cross-couplings within a layer and back-couplings to a previous layer. Transmission delays enable the perceptron to recognize temporal patterns. Numerous versions of cellular reaction types and of synaptic plasticity procedures are discussed. The book presents extensive statistical evaluations of cell numbers and of overlaps between sets of cells activated in the A-layers in response to different stimuli. A great number of experiments with concrete versions of perceptrons are presented and discussed. The parts most valuable to the present-day reader are, however, the various summaries and discussions at the beginnings and the ends of chapters. It is these which contain a wealth of ideas and insights regarding perception and brain function. Nobody was more aware of the limations of the perceptron approach than Rosenblatt, and one could only wish that more of the amateurs of brain theory read the book before they embarked on extensive projects in pattern recognition and brain modelling.

Rosenblatt's book crystallizes a number of concepts which together characterize the brain theory era of the last 25 years. Let me mention some of the most important of these concepts. Perceptrons are input-driven deterministic machines. A few transmission-delays after the presentation of a stimulus the machine has settled into a stable state. Neither has the perceptron spontaneity (e.g. active generation and testing of hypotheses), nor can it iterate more than a few times. This simplicity is dictated mainly by limitations in the power of analysis. No scheme establishing a connection between local structure and global function in back-coupled networks was known at the time.

The perceptron explicitly formulates a theory of perception. According to this, perception is a passive process (as has just been stated). The perceptron is based on the idea that a pattern on a sensory surface can be completely represented by the vector of activities of a fixed set of feature detector cells. Recognition of a pattern is tantamount to the activation of an R-cell ("pontifical cell"). A whole scene is represented by a set of such recognition cells. The perceptron thus explicitely formulates the position of semantic atomism, which to the present day is the basis of most of brain-theory: complex symbols (for the representation of patterns, scenes, etc.) are additively composed of atoms (cells), each of which has some more or less elementary bit of meaning attached to it (blue light on point x of retina, my grandmother, ...).

The perceptron exemplifies hetero-organization. The structure of a perceptron comes from two sources. It is "genetic" information which defines the functional constraints of cells and synapses, the division into planes (S, $A^{(1)}$, $A^{(2)}$, R), and especially also the structure of the $S \rightarrow A$ connections which determine the features that can be detected. The patterns to be recognized are defined by the "teacher", and are written into the system with the help of synaptic plasticity. The perceptron cannot develop new features or new patterns on its own.

The perceptron has various limitations, many of which have been discussed by Rosenblatt himself in his book. The most famous limitation of the three-layer perceptron is its unability to distinguish connected from unconnected figures. (This limitation is already mentioned in Rosenblatt's book and has been extensively discussed in a book by M. Minsky and S. Papert.) The difficulty can be easily mended in a scheme involving a miminum of serial computation.

A more serious limitation is the complete lack of syntactical information in the perceptron, as was pointed out by Rosenblatt himself. He employed the following example. Suppose there are four R-cells: $R_\triangle$ and $R_\square$ recognize a triangle and a square, $R_\uparrow$ and $R_\downarrow$ recognize that there is some pattern in the upper and in the lower half of retina. What, if there is a triangle in the upper half and a square in the lower half? All four cells go on. The situation is confused with one in which the positions of triangle and square are exchanged. Additional machinery for the expression of syntactical information would enable the system to make statements like "$R_\triangle$ and $R_\uparrow$ refer to the same object", and thereby resolve the ambiguity. In layer $A^{(2)}$ of a four-layer perceptron the same difficulty leads to more fundamental problems. There is no information in $A^{(2)}$ expressing the spatial grouping of features within the original pattern. This leads to the serious ambiguity between different patterns which contain the same local features but in a different arrangement. Also, information on the segmentation of a scene into objects cannot be expressed. This is a fundamental flaw of semantic atomism. Rosenblatt attempts a solution, by introducing extra connections in his perceptron with the result that activity in the R-layer is restricted to those cells which refer to one object only (e.g. $R_\triangle$ and $R_\uparrow$ in the example).

The whole perceptron approach has been haunted by a lack of appreciation of quantitative limitations. The convergence theorem, for instance, is of no practical use if convergence takes eons. Another limitation concerns numbers of feature detectors (in the A-layers). Pattern separation is only possible if the sets of activated feature cells are sufficiently disjunct. This in turn means that features must be rather specific to small pattern classes. In a rich environment the number of feature cells required diverges quickly. Realistic cell numbers rigidly limit the perceptron to very small pattern universes. A solution to this problem can only come from a scheme by which highly specific feature units are combinatorially built from low-specificity cells.

The perceptron and the ideas set out in Rosenblatt's book were of great historical importance. After an era which had indulged in formulating and reformulating universality claims in splendid generality, the perceptron focussed attention on an important particular problem, pattern classification (although Rosenblatt couldn't resist including a useless universality claim in his book). There may have been several hundred groups all over the world experimenting with and theorizing about perceptron-like structures. The perceptron and the associative memory continue to be the dominating ideas of brain theory, although they both date back to before 1960. The perceptron may be considered the last major artificial intelligence project which was formulated in neural terms and which has had close interactions with experimental neurobiology. After the perceptron (and perhaps partly in reaction to the disappointment with the perceptron) neurobiology and artificial intelligence have slowly lost contact and have ceased to inspire each other. Only recent years have seen a cautious revival of this symbiosis, in the form of massively parallel models of image interpretation and low-level vision.

On the surface, the reaction of the academic community to the perceptron has not exactly been a warm-hearted welcome. Some well-published theories were clearly elaborate versions of the perceptron yet failed to give explicit reference to Rosenblatt. Some of the reactions, especially the book by Minsky and Papert, identified the perceptron with its first bridge-head, the "simple perceptron", and then rejected it as being insufficient (see above), instead of taking it as a road into a new unknown land, as had been the intention of Rosenblatt. As a consequence of this reaction, perceptron theory has not become a focus of intellectual activity, in spite of its great influence on the practical activities of model builders. There may have been several reasons for this reaction of the academic community – the early frustrations after the exaggerated initial claims, the close engagement of Rosenblatt with the military, the inadequacy of the mathematical methods at the time, and the absorption of artificial intelligence into computer science and the algorithmic approach.

Two developments of recent years may revive interest in artificial intelligence "neural style". With probabilistic methods ("annealing", Boltzmann-Gibbs-statistics) it has recently become possible to theoretically derive the global behaviour of cross-coupled nervous network models from knowledge about local structure. Now, one can deal with cooperative phenomena and pattern formation in iterative processes. The other development is that the classical approach of computer science – algorithmically controlled sequential machines – is beginning to feel its quantitative limits with respect to software volume and computing time. For the future, one may hope that more intellectual effort will be devoted to nervous networks, that is, to self-organizing massively parallel structures. This development would have to take up the thread were Rosenblatt has dropped it.

Reference

Rosenblatt F (1961) Principles of neurodynamics: Perceptrons and the theory of brain mechanism. Spartan Books. Washington, DC

Wilfrid Rall: Electrophysiology of a Dendritic Neuron Model

A. LANSNER[1]

The major part of Rall's work has been devoted to the task of formulating a mathematical model of the passive electrical properties of neurons, their so called "cable properties". This work is perhaps most correctly referred to the branch of brain theory that may be characterized as theoretical neurophysiology. Rall's ambition has been to contribute to a "mathematical neuron theory" which would allow us to better understand the integrative properties of synaptic action, the behavior of large populations of neurons and, eventually, even phenomena like learning.

In the early 1960's, when Rall first addressed this problem, the active properties of neural tissue had been intensely studied for at least a decade (e.g. the work of Hodgin and Huxley). Taking a different approach, Rall directed attention at the extensive and complex dendritic branching of neurons and the "dendritic dominance" displayed by many types of neurons, i.e. the large surface area of the dendritic membrane compared to the soma. He was convinced that this had significant functional implications and that the quite commonly held view of a the neuron as a threshold element with impulse generating properties and an outgoing axon was far too simplistic.

There was already at that time a considerable amount of quantitative data on membrane properties of different types of neurons to feed a more complex and realistic neuron model. Cable theory had been developed earlier for other purposes but it had only to a limited extent been applied in neural modeling work. Rall started to adapt this theory to the branching, finite length dendritic structures of neurons. To be able to come up with a useful model, he made some simplifying assumptions. The most important was that, unlike axonal membrane, the dendritic membrane was assumed to be completely passive and to be uniform over the entire soma-dendritic surface. Furthermore, the soma was treated as isopotential, and the dendritic tree as composed of a number of cylindrical trunks.

[1] Department of Numerical Analysis and Computing Science, The Royal Institute of Technology, Stockholm, Sweden

Brain Theory
Edited by G. Palm and A. Aertsen

The theory was formulated as a set of partial differential equations with time and space as variables and with parameters for membrane properties, such as e.g. specific resistance and capacitance together with the spatial configuration of the dendritic structures (Rall 1962). Rall also proposed the so called 2/3-rule stating roughly that, at a branching point, the 2/3-power of the diameter of the entering parent branch equals the sum of 2/3-powers of the leaving branches. If the branching pattern of a dendritic tree obeys this rule, the passive electrical properties of the tree are equivalent to those of a cylinder shaped dendritic tree ("equivalent cylinder"). If this relation holds, it greatly simplifies application of the model.

Rall investigated the properties of this model thoroughly by means of both mathematical analysis and computer simulations. In the simulations compartemental models composed of equivalent electrical circuits for pieces of isopotential nerve membrane were used. With this model, Rall could derive a number of interesting theoretical results (Rall 1967), for example on the effect of a current step applied to the soma, the shape of the synaptic potential, the synaptic intensity required to produce a certain soma potential, the effect of a steady hyperpolarizing current on EPSP shape, and the detectability at the soma of transient synaptic conductance changes at different soma-dendritic locations. Rall summarized one of his more important early results as follows (pp. 154–155 in Rall 1962):

"This suggests a need for a modification of the widely held belief that synaptic excitation delivered to the dendritic periphery could produce no significant passive electrotonic spread of membrane depolarization to the soma. On the contrary, because of the very large dendritic surface area and the high density of synaptic contacts over the entire soma-dendritic surface ..., dendritic synaptic activity would be expected to dominate the slowly changing background level of the neuron's excitatory state."

Rall also suggested a number of quantitative shape indices for synpatic potentials at the soma that could be used to judge the location of synaptic input on the neuron. He also observed that multiple synaptic input close to the soma could be mistaken for distant input due to the time dispersion between individual synaptic potentials. Furthermore, Rall proposed that the location of synapses should be measured in terms of electrotonic distance rather than just as displacement in micrometers. One of his hypotheses related to the mechanisms of learning and synaptic plasticity. It was based on the idea that changes in the thickness and length of dendritic spines could modulate synaptic strength and therefore might be a physiological substrate of learning.

In several papers Rall cooperated with experimentalists in order to check his theoretical results against experimental data (see e.g. Rall et al. 1967). For instance, he has engaged in studies of how the amplitude and shape of soma postsynaptic potential could be used to judge the location of synapses in the dendritic tree. This method is still used and by applying it one has been able to demonstrate, for instance, that different fiber projections often make synaptic contact at different locations on the dendritic processes of their target neurons.

Today, the theory of the passive electrical properties of neurons represents a set of comparatively solid models that are also to some extent amenable to experimental testing. They have been elaborated, extended and modified many times over the years. The amount of data on membrane properties etc. is still growing fast and the quality of data has also been improved. In one form or the other, the equations developed constitute a standard component in neural modeling type of work together with equations describing the active properties of neural tissue (see e.g., MacGregor and Lewis 1977, Rall 1977, Traub and Llinas 1979, Perkel and Mulloney 1978). Several of Rall's early proposals are still largely relevant. For instance, it remains plausible, that the central nervous system exploits the degrees of freedom provided by passive electrotonic spread of current to tune different types of nerve cells to their particular task. Rall's suggestion of the function of dendritic spines in learning also remains a viable hypothesis.

On the other hand, today it is clear from experimental observations that several of the simplifications underlying this model are actually violated by real neurons. For instance, the dendritic membrane of some types of neurons is known to have active properties. Rall stated explicitly that his model did not apply in this case. It has to be extended with equations describing such phenomena accurately. Another problem when applying these models is that branching patterns obeying the 2/3-rule are often poor approximations to those of real dendritic trees. But the theory can well be applied without this rule although it complicates treatment considerably.

One may conclude, that the work of formulating a mathematical neuron model is still far from finished. It is certainly still a highly relevant ambition to try to construct and study the properties of even better neuron models. This will certainly contribute to a further improved understanding of the extremely complex network of billions of interacting neurons that makes up our brain.

References

MacGregor RJ, Lewis ER (1977) Neural modeling. Plenum Press, New York

Perkel DH, Mulloney B (1978) Electrotonic properties of neurons: Steady-state compartemental model. J Neurophysiol 41(3):621–639

Rall W (1962) Electrophysiology of a dendritic neuron model. Biophys J 2:145–167

Rall W (1967) Distinguishing theoretical synaptic potentials computed for different somadendritic distributions of synaptic input. J Neurophysiol 30:1138–1168

Rall W (1977) Core conductor theory and cable properties of neurons. In: Handbook of physiology, vol I(1). Brookhart M, Mountcastle VB, Kandel ER (eds) Am Physiol Soc, pp 39–97

Rall W, Burke RE, Smith TG, Nelson PG, Frank K (1967) Dendritic location of synapses and possible mechanisms for the monosynaptic EPSP in motoneurons. J Neurophysiol 30:1169–1193

Traub RD, Llinas R (1979) Hippocampal pyramidal cells: Significance of dendritic ionic conductances for neuronal function and epileptogenesis. J Neurophysiol 42:476–495

David Marr: A Theory of the Cerebellar Cortex

A Model in Brain Theory for the "Galilean Combination of Simplification, Unification and Mathematization"

A.J. PELLIONISZ[1]

The brilliant but tragically short-lived – indeed, meteoric – career of David Marr was catapulted into orbit by his "Theory of the Cerebellar Cortex" (1969). One need not belabor the details of this well-publicized theory here (cf. Eccles 1973, Llinàs 1981, Ito 1984, Pellionisz 1985). Rather, three general observations on the theory will be briefly discussed.

First, the idea behind the theory was very uncomplicated. The chief neurons of the cerebellar cortex, the Purkinje cells, would "learn" to recognize a pattern over many of their input fibers (the parallel fibers) *if* such pattern would coincide with the deep depolarization of the Purkinje cell, evoked by the dense arbor of the climbing fiber which innervates it.

Second, a theory of the cerebellar *cortex,* by definition, focuses rather narrowly on a specific part of the cerebellar system, and is further focused on only one (hypothetical) facet of cerebellar function; plasticity. While a sharply focused view is characterized by dramatic highlights, it is probably impossible to understand the full role of the cerebellum in motor control without looking at it from a less spectacular but both broader and deeper perspective.

For example, from the viewpoint of the structure, cerebellar function may remain an enigma if the role of other cerebellar structures than only its cortex, e.g., the cerebellar *nuclei,* is not understood or specific cerebellar functions are not interpreted in the context of the structure of a whole sensorimotor system. Remember the old, hitherto unanswered question: "Can we make a real systems approach to cerebellar function without modeling the whole motor system?" (Arbib et al. 1968).

From a functional viewpoint, it is essential to point out that the cerebellum has been known for more than a century as the organ in the brain which acts as a motor coordinator (Flourens 1842); it is classic knowledge that its absence produces spatial and temporal dysmetria (Holmes 1939). True, the coordinator function of the CNS, just like any other CNS function, is endowed with a good degree of adaptability (cf. Llinàs and Pellionisz 1985). However, Marr's theory only provided a hypothesis for plasticity, and none for coordination –

[1] Department of Physiology and Biophysics, New York University, 550 First Avenue, New York, NY 10016, USA

Brain Theory
Edited by G. Palm and A. Aertsen

in effect assuming that motor learning was *the* function of the cerebellum. It was probably this limited character of the theory of cerebellar cortex that led Marr, shortly before his untimely death, to repudiate his theory as a utilizable explanation of what role the cerebellum plays in the motor system: "In my own case, the cerebellar study had two effects. On the one hand, it suggested that one could eventually hope to understand cortical structure in functional terms, and this was exciting. But at the same time the study has disappointed me, because even if the theory was correct, it did not much enlighten one about the motor system – it did not, for example, tell one how to go about programming a mechanical arm" (Marr 1982, p. 15).

Third, the idea has completed its course during the past two decades. Since the original concept of the theory by Brindley (1964), a rather large initial theoretical followership of the idea emerged; see Smolyaninov (1966), Eccles et al. (1967), Szentàgothai (1968), Grossberg (1969), Marr (1969), Albus (1971). Later, championed by the eminent experimentalist (see Eccles 1977), a campaign was launched to "prove" the theory. The early results, however, showed that "unfortunately, experimental testing of this hypothesis has failed to discover any significant modification even after some hundreds of parallel fibre-climbing fibre inputs to Purkinje cells with a 10-ms time discrimination (J.C. Eccles, D. Marr, N.H. Sabah, R.F. Schmidt, and H. Tabořikova, unpublished observations)" (Eccles 1973). Despite such discouraging results, merely because of the appeal of the theory, the campaign was enlisted by workers too numerous to name here. One and a half decades of intensive and devoted research later, two pieces of circumstantial "evidence" have emerged. After applying repeated parallel fiber – climbing fiber stimuli, one study reports a "depression" at the Purkinje cell level (Ito et al. 1982), while the other, based on a somewhat different experimental paradigm, shows an "enhancement" (Bloedel et al. 1983). These apparently inconsistent phenomena, neither of which lasts for more than a matter of hours, appear to be meager support for a theory which is based on the notion that the cerebellar cortex "learns" movements by permanently imprinting patterns into a set of connectivities of Purkinje neurons. At present, only one remaining major school of experimentation retains the "cerebellar learning paradigm" as its theoretical foundation.

Most certainly, it is admirably easy to put limiting qualifications on any major school of thought *from the retrospect of two decades.* Let us see, however, whether it is equally obvious to discern those qualities of Marr's theory of the cerebellar cortex, which created the spectacular success that his theory truly amounted to. For a definitive answer, it is tempting to accept the very conductive notion by Churchland (1986) – that the most critical test of a theory is whether it provides a "Galilean combination: the right sort of simplification, unification, and above all, a *mathematization.*" Indeed, Marr's theory on the cerebellar cortex amply met each of these criteria.

In simplification, Marr's cerebellar theory even surpassed the Galilean viewpoint (*mutatis mutandis*). Identifying the function of the cerebellum as a uniquely simple operation – learning parallel fiber patterns by the Purkinje cells – was not unlike naming a planet around which the planetary system rotates. However, while the extraordinary simplicity of an assumption has its virtues, i.e. instant acceptance and universal acclaim, history documents that such axioms, even when shown inappropriate, are extremely difficult to dislodge. To shift the center of attention from a planet (or from plasticity) to a Sun (or to coordination) may take an inordinate amount of both time and exasperation.

Unification was perhaps the strongest virtue of Marr's theory of the cerebellar cortex. As he reminisced, his theory "suggested that one could eventually hope to understand cortical structure in functional terms, and this was exciting" (Marr 1982). An understatement. Indeed, by showing how the structure explains the function of *one third of the brain,* it was almost an entry into the promised realm of the inner sanctum of brain sciences. This was a refreshing contrast to the state-of-the-art of neuronal modeling, in which skilled applied mathematicians accounted for phenomena on rather minute fragments of nervous subsystems. His attempt at a grand unification of cerebellar structure and function was almost unprecedented, since such synthesis had been tackled only once before, by the pioneering notion that parallel fiber excitation volleys, arriving at Purkinje cells at different distances, would make the cerebellar cortex serve "as a clock in the millisecond range" (Braitenberg 1967). That earliest unification of micromorphological and overall functional features (beyond the problem of the demonstrably too short span of such timing), however, was based only on fairly trivial mathematization, and lacked some further unifying powers that Marr so deftly wielded in his theory. Two of the most important mergers that he accomplished were as follows: (1) Marr unified the knowledge of an anatomically distinct neuronal *circuitry* with a hypothesis at a *synaptic* level (that of Hebb 1949, who postulated that alteration of synaptic efficacies might subserve learning paradigms). (2) Equally important, Marr's theory of the cerebellar cortex unified the interpretation of the function of a *neuronal system* with theories of *abstract automata,* in the sense that through his theory the cerebellum appeared as a neuronal embodiment of Rosenblatt's (1959) learning machine, the Perceptron. Thus, from a wider perspective, Marr's theory – especially as exploited by Albus (1971, 1981) – was an attempt at unifying the fields of *Cybernetics* (cf. Wiener 1948) with biological studies of the cerebellum, which is at the forefront of *Neuroscience* (cf. Eccles et al. 1967).

Lastly, the *mathematization* of Brindley's (1964) germinal descriptive notion by Marr's quantitative elaboration must be addressed. As one of the most successful modelist in biology of all times comments: "It is not that most neurobiologists do not have some general concept of what is going on. The trouble is that the concept is not precisely formulated" (Crick 1979). As for Brindley's idea; for lack of mathematization it was dormant for years before Marr (1969), Grossberg (1969) and Albus (1971) matched the *concept* with a suitable mathematical *formalism.* This quantum jump elevated the notion into the league of theories, since a measure of the maturity and rigor of a field of science is the extent to which mathematics is utilized. By the "theory of cerebellar cortex" we have been given an example how brain research must progress towards exact synthesis through the sort of mathematization of biological concepts as achieved, for example, by Helmholtz (1896), Hodgkin and Huxley (1952) or Marr (1969).

Considering the mix of extraordinary merit and ordinary fallibility, what remains for us from Marr's theory of the cerebellar cortex?

It all depends on who we are. Naturally, in the field of experimental investigations, work always continues to "prove" or "disprove" *all* of the available theories, like people continue to savour all wine, old or new. Basing an experiment on a vintage theory is still better than working with no theoretical basis at all. Experimentalists can select from whatever theory is available (for recent reviews, see Eccles 1977, Llinàs and Simpson 1981, Ito 1984, Pellionisz 1985 or Pellionisz and Llinàs 1985).

For brain theory, however, the issue is not whether to prove or disprove theories with facts, if the old saying (attributed to Einstein) that "facts do not kill theories" is true. More advanced theories, like superior products of evolution, automatically *take precedence* over less potent earlier models; they all linger along, their fate being determined only by the rule of "the survival of the fittest".

Finally, for brain theorists, Marr's model of the cerebellar cortex remains forever a model for simplicity, unification, and mathematization; a high standard that one strives to equal, and perhaps to surpass.

References

Albus J (1971) A theory of cerebellar function. Math Biosci 10:25–61

Albus JR (1981) Brains, behavior and robotics. McGraw-Hill, New York

Arbib MA, Franklin GF, Nilsson N (1968) Some ideas on information processing in the cerebellum. In: Caianiello ER (ed) Neuronal networks. Springer, Berlin Heidelberg New York, pp 43–58

Bloedel JR, Ebner TJ, Qi-Xiang Yu (1983) Increased responsiveness of Purkinje cells associated with climbing fiber inputs to neighboring neurons. J Neurophysiol 50:220–239

Braitenberg V (1967) Is the cerebellar cortex a biological clock in the millisecond range? In: Fox CA, Snider RS (eds) Progress in brain research, vol 25. The cerebellum. Elsevier, Amsterdam, pp 334–346

Brindley GS (1964) The use made by the cerebellum of the information that it receives from sense organs. IBRO Bull 3:80

Churchland PS (1986) Neurophilosophy: Towards a unified understanding of the mind-brain. Bradford Books/MIT Press, Boston, Massachusetts

Crick FHC (1979) Thinking about the brain. Sci Am 241:219–232

Eccles JC (1973) The cerebellum as a computer: Patterns in space and time. J Physiol London 229:1–32

Eccles JC (1977) An instruction-selection theory of learning in the cerebellar cortex. Brain Res 127:327–352

Eccles JC, Ito M, Szentàgothai J (1967) The cerebellum as a neuronal machine. Springer, Berlin Heidelberg New York

Flourens P (1842) Recherches experimentales sur les proprietes et les fonctions du systeme nerveux dans les animaux vertebres, 2nd edn. Bailliere

Grossberg S (1969) On learning of spatiotemporal patterns by networks with ordered sensory and motor components. 1. Excitatory components of the cerebellum. Stud Appl Math 48:105–132

Hebb DO (1949) The organization of behaviour. Wiley, New York
Helmholtz H (1896) Handbuch der Physiologischen Optik. Voss, Leipzig
Hodgkin AL, Huxley AR (1952) A quantitative description of membrane current and its application to conduction and excitation in nerve. J Physiol (London) 1176:500–544
Holmes G (1939) The cerebellum in man. Brain 63:1
Ito M (1984) The cerebellum and neural control. Raven Press, New York
Ito M, Sakurai MK, Tongroach P (1982) Climbing fibre induced depression of both mossy fibre responsiveness and glutamate sensitivity of cerebellar Purkinje cells. J Physiol (London) 324:113–134
Llinàs R (1981) Cerebellar modeling. Nature (London) 291:279–280
Llinàs R, Pellionisz A (1985) Cerebellar function and the adaptive feature of the central nervous system. In: Berthoz A, Melvill-Jones G (eds) Reviews of oculomotor research, vol I. Adaptive mechanisms in gaze control. Elsevier, Amsterdam, pp 223–231
Llinàs R, Simpson JI (1981) Cerebellar control of movement. In: Towe AL, Luschei ES (eds) Handbook of behavioral neurobiology, vol V. Motor coordination. Plenum Press, New York, pp 231–302
Marr D (1969) A theory of the cerebellar cortex. J Physiol (London) 202:437–470
Marr D (1982) Vision. A computational investigation into the human representation and processing of visual information. Freeman, San Francisco
Pellionisz A (1985) Tensorial brain theory in cerebellar modeling. In: Bloedel et al. (eds) Cerebellar functions. Springer, Berlin Heidelberg New York, pp 201–229
Pellionisz A, Llinàs R (1985) Tensor network theory of the metaorganization of functional geometries in the CNS. Neuroscience 16:245–273
Rosenblatt F (1959) Two theorems of statistical separability in the perceptron. Proceedings of Symposium on the mechanization of thought process. HMSO, London, pp 421–456
Smolyaninov VA (1966) Some special features of organization of the cerebellar cortex. In: Gelfand IM, Gurfinkel VS, Fomin SV, Tsetlin ML (eds) Models of the structural-functional organization of certain biological systems. MIT Press, Cambridge, pp 251–325 (Translated from Russian 1971)
Szentàgothai J (1968) Structuro-functional considerations of the cerebellar neuron network. Proc IEEE 56:960–968
Wiener N (1948) Cybernetics, or control and communication in the animal and the machine. MIT Press, Cambridge

David Marr: A Theory for Cerebral Neocortex

W. VON SEELEN[1]

This paper is an important contribution to the understanding of the visual system, it contains a part of those ideas which have become the commonly accepted basis of current research. Although some of these principles already had a history in 1970, Marr clearly deserves the credit for their sharp formulation and for a series of attempts leading to a formalization of the problems. His way of dividing the approach into the levels of computational theory, of the algorithm and of the implementation clarified the problems. His creed that human visual processing is modular, and that different types of information, which are encoded in the image can be decoded independently by modules, has been generally accepted. His idea that the neuronal structure is adapted to the structure of the signals from the environment, which can be viewed as locally homogeneous, constant and with a tendency towards continuity, has given rise to fruitful research. The idea of concept formation occupies an important place. This principle holds that the diagnosis of the outer world is realized on the basis of forming concepts (chair, lover, music etc.), and that basic concepts form a hierarchy with different levels of abstraction. Although this idea is older than his paper, his formalization of it in statistical terms is one of the high points of the paper along with his statements on redundancy.

The proposed neuronal realizations of the concepts discussed are partly dependent on knowledge available at that time, and some have been superseded by later findings. The same holds for the strategy for forming concepts and constructing hierarchies. However, current ideas on these topics are also only hypotheses. Whatever objections may be raised to details, the paper is an extraordinarily intelligent outline, which has impressed its mark on many later studies. Reading it today is still very stimulating and highly to be recommended.

Reference

Marr D (1970) A theory for cerebral neocortex. Proc R Soc London Ser B176:161–234

[1] Institut für Zoologie III (Biophysik), Johannes Gutenberg Universität, Saarstraße 21, 6500 Mainz, FRG

Brain Theory
Edited by G. Palm and A. Aertsen

MIX
Papier aus verantwortungsvollen Quellen
Paper from responsible sources
FSC® C105338

If you have any concerns about our products,
you can contact us on
ProductSafety@springernature.com

In case Publisher is established outside the EU,
the EU authorized representative is:
Springer Nature Customer Service Center GmbH
Europaplatz 3, 69115 Heidelberg, Germany

Printed by Libri Plureos GmbH
in Hamburg, Germany